"This important book is an unusually topical attempt to introduce readers to the relationship between the technical analysis of financial market prices and the automated implementation of its findings. The book will be of considerable interest to those who wish to know about this relationship in an eminently readable form: both professional financial market analysts and those considering future employment in the field."

—**Michael Dempster**, Professor Emeritus in the Statistical
Laboratory at the University of Cambridge

"AI is an important part of finance today. Students who want to join the finance industry should read this book. The trained eyes will also find a lot of insights in the book. I cannot think of any other book that teaches computational finance at a beginner's level but at the same time is useful to practitioners."

—**Amadeo Alentorn**, PhD, Head of Systematic
Equities at Jupiter Asset Management

"**AI for Finance** is an excellent primer for experts and newcomers seeking to unlock the potential of AI. The book combines deep thinking with a bird's eye view of the whole field - the ideal text to get inspired and apply AI. A big thank you to Edward Tsang, a pioneer of AI and quantitative finance, for making the concepts and usage of AI easily accessible to academics and practitioners."

—**Richard Olsen**, Founder and CEO of Lykke, co-founder of OANDA,
and pioneer in high frequency finance and fintech

"Without a doubt, AI symbolizes the future of finance and, in this important book, Professor Tsang provides an excellent account of its mechanics, concepts and strategies. Books featuring AI in finance are rare so practitioners and students would do well to read it to gain focus and valuable insights into this fast-evolving technology. Congratulations to Professor Tsang for providing a readable and engaging work in a complex technology that will appeal to all levels of readers!"

—**Dr David Norman**, *Founder of the TTC Institute*

"The use of AI/ML in the financial industry is now more than a hype. In financial institutions there are numerous active transformation programs to introduce AI/ML enabled products in areas such as risk, trading and advanced analytics. In this book, Edward, one of the early adopters of AI in finance, has provided an insightful guide for both finance practitioners and academics. I can see this book becoming a major reference in real-world applied AI in finance. Directional Change (Chapter 6) should be of particular interest to data scientists in finance, as how one collects data determines what one can reason about."

—**Dr Ali Rais Shaghaghi**, *Lead Data Scientist at NatWest Group*

AI for Finance

Finance students and practitioners may ask: can machines learn everything? Could AI help me? Computing students or practitioners may ask: which of my skills could contribute to finance? Where in finance should I pay attention? This book aims to answer these questions. No prior knowledge is expected in AI or finance.

To finance students and practitioners, this book will explain the promise of AI, as well as its limitations. It will cover knowledge representation, modelling, simulation and machine learning, explaining the principles of how they work.

To computing students and practitioners, this book will introduce the financial applications in which AI has made an impact. This includes algorithmic trading, forecasting, risk analysis portfolio optimization and other less well-known areas in finance.

This book trades depth for readability. It aims to help readers to decide whether to invest more time into the subject.

This book contains original research. For example, it explains the impact of ignoring computation in classical economics. It explains the relationship between computing and finance and points out potential misunderstandings between economists and computer scientists. The book also introduces Directional Change and explains how this can be used.

Edward P. K. Tsang

Edward Tsang is a retired professor and a freelance consultant. With a first degree in finance and a PhD in AI, he has broad interests in constraint satisfaction, optimization, AI and finance.

AI FOR EVERYTHING

Artificial intelligence (AI) is all around us. From driverless cars to game winning computers to fraud protection, AI is already involved in many aspects of life, and its impact will only continue to grow in future. Many of the world's most valuable companies are investing heavily in AI research and development, and not a day goes by without news of cutting-edge breakthroughs in AI and robotics.

The *AI for Everything* series explores the role of AI in contemporary life, from cars and aircraft to medicine, education, fashion and beyond. Concise and accessible, each book is written by an expert in the field and will bring the study and reality of AI to a broad readership including interested professionals, students, researchers, and lay readers.

For more information about this series please visit:
https://www.routledge.com/AI-for-Everything/book-series/AIFE

AI FOR FINANCE

Edward P. K. Tsang

CRC Press
Taylor & Francis Group
Boca Raton London New York

CRC Press is an imprint of the
Taylor & Francis Group, an **informa** business

First edition published 2023
by CRC Press
6000 Broken Sound Parkway NW, Suite 300, Boca Raton, FL 33487-2742

and by CRC Press
4 Park Square, Milton Park, Abingdon, Oxon, OX14 4RN

CRC Press is an imprint of Taylor & Francis Group, LLC

Library of Congress Cataloging–in–Publication Data

Names: Tsang, Edward, author.
Title: AI for finance / Edward P.K. Tsang.
Description: First edition. | Boca Raton : CRC Press, 2023. | Series: AI for everything | Includes bibliographical references and index.
Identifiers: LCCN 2022055176 (print) | LCCN 2022055177 (ebook) | ISBN 9781032391205 (hardback) | ISBN 9781032384436 (paperback) | ISBN 9781003348474 (ebook)
Subjects: LCSH: Investments--Data processing. | Finance--Data processing. | Artificial intelligence--Financial applications.
Classification: LCC HG4515.5 .T77 2023 (print) | LCC HG4515.5 (ebook) | DDC 332.0285/63--dc23/eng/20221119
LC record available at https://lccn.loc.gov/2022055176
LC ebook record available at https://lccn.loc.gov/2022055177

ISBN: 9781032391205 (hbk)
ISBN: 9781032384436 (pbk)
ISBN: 9781003348474 (ebk)

DOI: 10.1201/9781003348474

Typeset in Joanna
by Deanta Global Publishing Services, Chennai, India

CONTENTS

To my parents

ACKNOWLEDGEMENTS

I am indebted to Jim Doran, who introduced AI to me nearly 40 years ago and guided me through my PhD in AI.

Without teaching from Richard Olsen, Raju Chinthalapati and David Norman, my knowledge of computational finance would have been shamefully shallow.

I would like to thank my co-researchers, Shu-heng Chen, Michael Kampouridis, James Butler, Yi Cao, Tun-Leng Lau, Jin Li, Mathias Kern, Nanlin Jin, Serafin Martinez-Jaramillo, Biliana Alexandrova-Kabadjova, Alma Lilia Garcia-Almanza, Amadeo Alentorn, Ali Rais Shaghaghi, Abdullah Alsheddy, Shaimaa Masry, Monira Al Oud, Han Ao, Jorge Faleiro, Amer Bakhach, Antoaneta Serguieva, Jun Chen, Shui Ma, Shengnan Li, Shicheng Hu and Sara Colquhoun.

I am also grateful to my ex-colleagues, Michael Dempster, Sheri Markose, Dietmar Maringer, the late Nick Constantinou, John O'Hara, Wing Lon Ng, Steve Phelps, Abhinay Mutoo, Kyriakos Chourdakis, Neil Kellard, Qingfu Zhang, John Ford, Ray Turner, Richard Bartle, Libor Spacek, Jeff Reynolds, Paul Scott, Sam Steel, Chang Wang, Hani Hagras, Lyudmyla Hvozdyk, Elena Medova, Anne De Roeck, Massimo Poesio and Michael Fairbank, for educating me on many aspects of computation and finance.

Much of my knowledge about the finance industry also comes from Carlo Acerbi, Laurence Wormald, Amadeo Alentorn, Angelo De Pol, Giovanni Beliossi, Paul Ingram, Shane Lamont, Evi Pliota, Rafael Velasco-Fuentes, Gaelle De Sola, Rhomaios Ram, Paula Haynes, Carlo Rosa, Tim Clarke, Bob Berry, Manfred Gilli, Philip Treleaven, Willem Buiter and many others.

Ideas in this book matured through my interacting with students at the University of Essex, King's College London and the University of Hong Kong, to whom I am grateful. Fellow researchers in conferences and seminars, especially members of the IEEE Technical Committee for Computational Finance and Economics, have all contributed to the development of my ideas in this book. I hope they will forgive me for not putting every name down; their support has not been forgotten.

Finally, I must thank Elliott Morsia; without his encouragement, this book may never have been written.

PREFACE

I have witnessed remarkable developments in computational finance. I would like to believe that I and my team have helped its development. I want to write a book to summarize some of the developments that are close to my research. I have been planning to write a book on this topic for over ten years. But this book was not what I had in mind.

Before I started writing this book, I was preparing a more serious book on the same topic. I thought this book would be a distraction from the serious book. As I started writing this book, I realized the value of writing it: I cannot express my opinions freely in a more serious book – I must substantiate every point and carefully provide references. As a leisure read, this book allows me to express my opinions more freely. In fact, I hope my opinions are valuable contributions to the field. I hope they provide insight and provoke discussions. For example: where could I say "*between classical economics and AI, neither can live while the other survives*" (Chapter 2) in a serious text?

Wait a minute. Am I suggesting that this is not a serious book? While this book may not be publishable as an academic text, it IS serious! The topic is serious. This is a serious attempt to explain complex material to the public. Reporting 30 years of research in a short text requires a lot of serious work. I have seriously enjoyed writing this book. I hope this book provides the readers with some serious fun!

Edward Tsang
November 2022

INTRODUCTION

"This is the best of times; this is the worst of times".[1]

This is the best of times; this is the worst of times for fund managers, traders and investors. This is the best of times for those who manage to take advantage of advanced technology, AI being an important part. They are armed with advanced weaponry and therefore have a better chance of survival. This is the worst of times for those who are left behind. If they fail to appreciate what their opponents are using, they have no chance to survive in financial markets.

This book aims to explain (1) from the business point of view, where finance could benefit from AI and (2) from the technology point of view, where AI could contribute to finance.

From the business point of view, AI has rewritten the finance industry for those who have already used AI (those who haven't been left behind). This book will identify some of the finance operations that could take advantage of research in AI. This includes algorithmic trading, forecasting (Chapter 3), risk analysis (Chapter 4), portfolio optimization (Chapter 5) and data handling (Chapter 6).

From the technology side, Google's AlphaGo brought machine learning to many people's attention. Can machines learn everything? This book explains the basic principles of machine learning and their limitations (Chapter 2). It also explains some of the AI techniques that financial experts could adopt to gain an edge over their competitors who have failed to take advantage of AI developments.

This book starts by explaining the synergy between AI and finance: it highlights how financial knowledge and AI knowledge could work together to achieve what neither a finance expert nor an AI expert could achieve on their own (Chapter 1).

Google's stunning success in the boardgame Go ignited worldwide enthusiasm for machine learning. It led some to believe that machines can learn anything by themselves. Chapter 2 explains the promise of machine learning and the limitation of computation in general. Readers need to understand such limitations and how they impact the basic assumptions of classical economics.

Chapter 3 explains how machine learning works. With the help of two financial applications (forecasting and bargaining, a branch of game theory), it explains two classes of machine learning, namely "supervised" and "unsupervised" learning.

Chapter 4 explains how modelling, simulation and machine learning could be combined to form a powerful tool in finance. This is illustrated in risk assessment, trading strategies design and the design of rules in new markets.

Chapter 5 introduces the portfolio optimization problem. It explains the current practices and their limitations. It explains that researchers and practitioners are hardly addressing the real problem, hence opportunities lie ahead.

This book encourages readers to think outside the box: AI is not just about algorithms; knowledge representation is an important part of early AI. Financial researchers and practitioners are all familiar with Time Series. Is Time Series the most natural way to represent time? What is time anyway? This will be explored in Chapter 6.

Chapter 7 briefly covers some of the research that we cannot fit into this short book: algorithmic trading has become a machines-deceiving-machines battle. The significance of high-frequency finance is explained. Then it explains that blockchain could provide a lot of opportunities for AI. It also explains that information does not only come from market data, but it may also come from news and social media too. Finally, it explains why opportunities are abundant in today's markets.

Table 0.1 summarizes the financial topics covered in this book and the computational research which are relevant to those topics.

This book assumes knowledge in neither AI nor finance by its readers. It could be read by anyone with a general interest in finance and AI. Having said that, readers with AI or finance knowledge will gain a deeper understanding of the material. The Bibliographical Remarks section provides pointers to publications for those who are keen to find out more about specific topics.

Table 0.1 Financial Topics Covered in this Book and the Supporting Technology

Topics in finance	Supporting technology
Algorithmic trading (Chapter 1, Section 7.1)	Machine learning (Chapter 2)
Forecasting (Section 3.1)	Supervised learning (Section 3.2)
Bargaining in game theory (Section 3.4)	Unsupervised learning (Section 3.5)
Risk analysis (Section 4.4)	Modelling and simulation (Section 4.4)
Payment systems (Section 4.1)	Modelling (Chapter 4)
Trading strategies design (Section 4.5)	Unsupervised learning (Section 4.5)
Mechanism design (Section 4.6)	Modelling, simulation and learning (Section 4.6)
Portfolio optimization (Chapter 5)	Optimization (Sections 5.1, 5.2), Constrained optimization (Section 5.3), Multi-objective optimization (Section 5.4)
Directional Change (Chapter 6)	Knowledge representation (Chapter 6)
High-frequency finance (Section 7.2)	Directional change (Chapter 6), Novel methods required

NOTE

1. Adopted from Charles Dickens, *A Tale of Two Cities*, 1859.

1

AI-FINANCE SYNERGY

From the business point of view, where could finance benefit from AI? If I am a trader, a fund manager or a risk manager, how can I use AI techniques to gain an edge over my competitors? Is AI all about speeding up computation?

From the technology point of view, where could AI contribute to finance and how? If I am an AI expert, where can I contribute my knowledge to financial institutes? If I am a student wanting to join the finance industry, what AI techniques should I pay attention to? Is it straightforward to apply AI techniques to finance?

How is big data related to AI and finance? Can machines learn everything by themselves? If so, can AI replace human experts in finance?

This book aims to answer the above questions. In this chapter, we shall use algorithmic trading as an example to illustrate how machines could work with people.

1.1 SPEED MATTERS

Let us start with a tale of two city traders, Alan and Beatrice.

Human traders attempt to valuate assets. Traders buy if the price of an asset is lower than their valuation and sell when the price is above their valuation. They do so because they believe that the

DOI: 10.1201/9781003348474-1

price will reflect the asset's true valuation in the long term. In other words, they predict that price movement will agree with their valuation. They make money if the market does agree with their valuations after they have bought or sold, or if they are lucky – but this book is not about gambling.

Suppose trader Alan is right in predicting that the price of an asset will rise. He must act fast because when other traders see the same, they will start buying, which will push the price up. In other words, being able to predict price movement is not enough, a trader must buy ahead of others.

That is where computers come in. The simplest form of algorithmic trading is to program a trader's trading strategy into a computer. The program will then act on behalf of the trader. The data feed will enable the computer program to monitor the market and buy and sell when the specified conditions are met.

Here, in the simplest form of algorithmic trading, no AI is involved. The program simply implements a trader's strategy. Trader Alan benefits from (a) his trading strategy and (b) the speed of computers and networks.

Suppose trader Beatrice uses exactly the same trading strategy as Alan. To beat Alan, Beatrice could implement this strategy on a faster computer. Alternatively, Beatrice could invest in faster networks. By doing so, Beatrice will beat Alan in placing her order ahead of Alan's. By the time Alan places his order, the price could have already gone up.

In this scenario, speed matters. The winner is the one who has faster programs, faster computers, faster data feed, faster network or is able to place their orders faster than their opponents.

1.2 THE RACE IS ON SEEKING, NOT RUNNING

Speed matters in the above scenario. Late comers can only trade at the prices that they want if the winner has not moved the price.

The assumption above is that both traders trade with the same strategy. This will happen if they both use textbook strategies, or they can see the same obvious, risk-free opportunities in the market. The question is: how often are obvious opportunities available?

How often does one see a piece of gold lying there to be picked?

Risk-free trading opportunities should not exist, theoretically, but they do; rare though they are. Following are two examples:

Exchange Rate Discrepancy:

Suppose 1 Euro buys US$1.158, which buys £0.86. If an exchange company offers to buy 1 Euro for £0.87, then a trader can generate a profit of £0.01 through each cycle of selling Euro for US$, selling US$ for £ and selling £ for US$. This opportunity will last before the exchange rates change.

Arbitrage in Derivatives:

Futures[1] and options[2] are derivatives of assets. The prices of futures and options of HSBC, for example, must be related to the price of HSBC shares. Occasionally misalignments do happen. When such opportunities arise, traders can set up "arbitrage" contracts to buy the underpriced derivatives and sell the overpriced derivatives simultaneously to make a profit. As transactions change prices, the opportunities disappear as trades happen, hence speed is important.

Asset mispricing is not as uncommon as one might expect. Evidence suggests that arbitrage opportunities described above did exist. Well-known opportunities like the above are exploited by big players who can afford the fastest machines and networks. However, competition in speed is costly as equipment becomes outdated quickly. As these opportunities are well known, big players may join the race at any time. That makes winning harder, costly and not guaranteed.

As competing in speed is costly without a guarantee of success, it is more cost-effective to invest one's effort in discovering new opportunities instead. The deeper an opportunity hides, the more valuable they are because fewer people will be able to find them. So, it is worth searching deeper into data for regularities, which may be translated into opportunities. If one manages to discover new opportunities ahead of others, one could exploit them before others. Therefore, the real competition is:

How does one find opportunities ahead of others?

1.3 PATTERN RECOGNITION

A branch of financial analysis is called technical analysis. They believe that while individual traders' decisions may not be predictable, collective behaviour can often be observed. For example, when two people enter a lift, it is likely that they will occupy opposite corners. They believe that while the true values of assets matter in the long term, traders react upon the immediate price movements. Such reactions form patterns. Patterns may disappear when enough traders act upon them. If a trader can recognize such patterns before they disappear, profits can be made.

Technical analysts believe that a substantial number of patterns have been found in many markets. Fundamental analysts, on the other hand, believe that patterns like this are accidental. They believe that all information about an asset, including any patterns found, is reflected in the price of the asset, and therefore cannot be exploited for profit.

We are not going to join the debate of whether technical analysis is sound or not; a huge amount of research has been published on this topic. What matters here is that traders who believe in technical analysis will trade with rules derived from such analysis. AI techniques can help them to find such rules.

To help our discussion, we shall introduce the idea of moving average (MA), which traders use to capture the momentum of the market. For example, the 7-day MA computes the average prices of an asset in the past 7 days. The 21-day MA computes the average prices in the past 21 days.

If the 7-day MA was lower than the 21-day MA yesterday, but higher than the 21-day MA today, then we say that the market momentum is on the rise. Here 7 and 21 are just used as examples. In general, if the short-term MA crosses the long-term MA from below, then the prices are said to be rising. On the other hand, if the short-term MA crosses the long-term MA from above, then we say that prices are said to be falling. Based on this belief, two momentum trading strategies can be defined.

Momentum Trading Strategies
A **Trend Follower** will buy when the short-term MA crosses the long-term MA from below; sell when the short-term MA crosses the long-term MA from above.

A **Contrarian** will buy when the momentum shows the price is falling and buy when it is on the rise.

It is worth reiterating that we are not taking a position on the debate of whether such trading strategies have a sound foundation or not. They are important as long as they are popular.

The important point is that once a pattern is well known, it will be exploited by traders. When several traders use the same strategy, the competition reverts to the hardware race. Once the fastest traders trade, profiting opportunities will cease to exist. Hence, momentum strategies will no longer be reliable for other traders.

In general, simple patterns will be found quickly by many traders. To stay ahead of competitors, a trader must keep finding new patterns all the time. As the Red Queen said:

It takes all the running you can do, to keep in the same place.[3]

1.4 DATA MINING

For traders who do not want to participate in the expensive hardware competition, seeking new patterns remains the best strategy. For them, the competition is in the speed of seeking, not in the speed of hardware. Those who manage to discover new patterns fast than their competitors will win the competition. As we shall explain, those who know how to use AI techniques stand a better chance of discovering new patterns ahead of others.

What kinds of regularities could one possibly find? Technical analysis focuses on price movements alone. A deeper analysis will be able to find regularities based on economic and financial foundations. For example, changes in interest rates affect bond and stock prices. One may be able to find regularities between the economic climate and oil prices. Besides, an individual stock's price is not only affected by the prospect of a company. It is also affected by the overall mood in the market. Currency exchange rates affect a country's economy, so as the unemployment rate, consumer price index, industrial producer price index, etc. All these factors affect the price of an asset. Finding the regularities among all these factors may not be possible. But one may benefit from being able to find partial regularities among some of these factors.

Examining possible regularities between every combination of assets, exchange rates and economic indicators is out of the question, due to the sheer number of combinations. In fact, examining one single combination of two objects can be laborious. For example, when the Federal Reserve Bank changes the interest rate, how would the US dollar to euro (USD-EUR) exchange rate be affected? Would the effect be instantaneous? If not, how long would the effect last? Is the relationship linear? That means, does raising the interest rate by 0.5% have double the effect of raising it by 0.25%? This is where machine learning comes in. Given that this is a hide-and-seek competition, machine learning will help a seeker to find patterns, if they exist, ahead of its competitors. Machine learning will be explained later in the book.

Machine learning helps, but knowing where to look is more important. In the above example, we know that interest rate changes by the Federal Reserve Bank affect the USD-EUR exchange rate, but there is no guarantee that we can discover the exact regularities. It is likely that other facts must be considered. The inflation rate? The unemployment rate? The oil prices? Where should one start looking? What combination is more promising? Seeking a needle in a haystack blindly is unlikely to succeed. In the competition of seeking, knowing where to look first gains the seeker an edge over one's competitors. This is where financial expertise comes in.

One must work hard to find more and more complex patterns. To do that, the search must be guided by financial experts. They are in a position to tell where promising areas are. But to find complex patterns, one must search deeply into regularities. For that, human effort is not efficient enough. Even if they can discover patterns, they may not be able to discover them ahead of their competitors. Machine learning may help.

Figure 1.1 shows how machine learning could be used in algorithmic trading. The financial experts will identify a set of variables which they consider relevant to the trading of an asset that they are interested in. Historical data in these variables are fed into a machine learning system which will attempt to find regularities. The machine learning system could learn from historical data how it could trade for profit. It could learn the conditions under which it should buy, hold or sell an asset. It could also learn how to manage risk and manage cash flow. The goal of the machine learning system in Figure 1.1 is to generate a trading program.

Once training is complete, unseen data will be streamed into the trading program. It will constantly check whether the trading conditions are met. If they are, the program can trade autonomously. The hope is that the conditions that worked for historical data will work on unseen data. The learning process may take time. The trading system will be able to react to the input data within milliseconds. More on algorithmic trading will be discussed in Section 7.1.

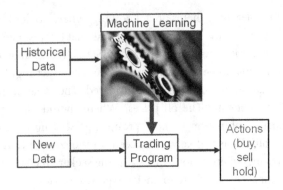

Figure 1.1 From Data to Algorithmic Trading through Machine Learning.

1.5 FORECASTING

Algorithmic trading is just one of the many things that machine learning could be applied to. In this section, we shall look at two other examples of machine learning.

One popular branch of research is *forecasting*. Asking different questions has different implications. For example, one might ask:

Question 1: *"What will the FTSE 100 Index be tomorrow?"*

Alternatively, one could ask:

Question 2: *"Will the FTSE 100 Index rise by 4% within the next 7 days?"*

Answering the first question is harder than answering the second question. Answering these two questions demands different techniques. The answer to Question 1 is probably more useful than the answer to Question 2, but a trader will be able to benefit from the answer to either question. What the trader is more concerned about is how accurate the forecasts are.

From a machine learning expert's point of view, the two questions demand different techniques. Question 1 demands an answer to a number. Therefore, the machine learning technique should be quantitative in nature. It may target to find a mathematical function that relates the numerical values of some indicators to the FTSE 100 index. Question 2 demands a Boolean – "yes" or "no" – answer. Learning some mathematical functions, as in Question 1, could help; but it may also be helped by a machine learning method that learns some logical relationships. It takes good knowledge of machine learning to know which methods suit which problem.

One popular approach to answer Question 1 is by inputting into the machine learning program the past, say, 50 days' prices. Fundamental analysts (introduced in Section 1.3) would believe that this is futile, as the current price will reflect all the information contained in those 50 days' prices. Even if the technical analyst were right that trading patterns could be formed in the market, simple patterns will be detected in this simple approach. To find patterns that others have not yet discovered, a trader must try harder. To try harder, one may input to the machine learning system factors that are not used by others. Alternatively, special methods must be developed to search for information buried in deep structures in the data – arguably this is much harder because of the sheer number of researchers who have tried it. The ones who succeed tend to be those who use both financial and machine learning expertise.

Forecasting does not have to be perfect. One does not have to correctly forecast every time. If a forecast is good enough to turn a 50–50 chance to 60–40 in the trader's favour, then it is potentially very useful.

One popular trading strategy that plays on the odds is *statistical arbitrage*. The program attempts to identify two assets whose prices historically move together, and trade when their movements deviate from the norm. For example, input to a machine learning system is the 5 years' daily closing prices of the S&P500 stocks. The system will compute the cointegration between pairwise stocks. Roughly speaking, this means finding pairs of stocks whose price differences tend to lie within a small range. For a pair of stocks that are found to be highly cointegrated, the program determines the normal range within which their prices move together. When their price difference deviates from the normal range by a big enough margin, the program will short one stock[4] and long[5] the other.

Back to forecasting, it is worth mentioning that even when a dependency relationship is observed between the input and the target variables, the relationship may not persist. The relationship observed may be born out of coincidence. One such example is the Lipstick Index. At some point, there was evidence showing that lipstick sales increased during a poor economy. This phenomenon ceased to be true in later recessions, which debunked the Lipstick Index.

Finally, it is worth pointing out that forecasting alone is not enough. To turn a good forecast into trading strategies, money management must be added: when the price is forecasted to rise, how much capital should the trader commit to the asset? When the price has risen as forecasted, should the trader take some profit? Machine learning can be used to examine the effect of different trading strategies, but ultimately, financial expertise is required to lead the research.

1.6 CONCLUDING SUMMARY: SYNERGY BETWEEN AI AND FINANCE

To summarize, when obvious opportunities are observable in the market, trading is a competition in speed – network, software and hardware speed all matter. However, obvious opportunities will be exploited quickly by those who invest in the fastest computers and networks. Thus, competition in computing speed is costly. A better way to protect one's investment is to research to discover opportunities ahead of one's competitors. The speed in seeking opportunities, as opposed to the speed of computing equipment, is where most of the competition lies.

In the opportunities-seeking competition, expertise in both finance and computing matters. Financial experts know where to look; computing experts can search fast. Together they stand a good chance of finding exploitable regularities ahead of financial experts or computing experts who work without the other expertise. This will be elaborated in Chapter 2.

NOTES

1. Futures are the obligation to buy or sell at a certain price at a fixed time in the future.

2. Options are the right to buy at a certain price at a fixed time in the future.

3. Borrowed from Lewis Carroll, *Through the Looking-Glass*, Macmillan, 1872.

4. Shorting an asset means selling the asset when a trader does not own any of it. The trader normally has to pay to borrow the asset from someone who holds it.

5. Longing an asset means starting with the position of holding none of the asset, buying some of it.

2

MACHINE LEARNING KNOWS
NO BOUNDARIES?

2.1 ALPHAGO: THE SUCCESS

AlphaGo, now owned by Google DeepMind, is a computer program that plays the boardgame Go. Go is a two-player game. The board is made up of a grid that comprises 19 horizontal lines and 19 vertical lines. The two players take turns to place pieces on the board. One piece is placed per turn on one of the unoccupied positions in the 19 × 19 grid. The goal is to occupy more territory than the opponent at the end of the game.

The rules of the game are unimportant to our discussion here. The important thing to know is that the first player has a choice of 381 (19 × 19) positions to place their piece. The second player has 380 remaining positions left to place their piece. Without complicating the discussion here, we can assume that the game could end in one of the 381 factorial possible sequences.[1] That is more sequences than the number of molecules in the universe. The implication of this is that even the fastest computer in the world today has no chance of evaluating all possible sequences of the game within one's lifetime.

In 2016, AlphaGo beat the professional human player Lee Sedol 4—1 in a five-game match. It went on to beat Ke Jie, the number one world-ranking player at the time, 3—0 in a match in 2017. These

DOI: 10.1201/9781003348474-2

results stunned the world, as nobody at the time expected computer programs to beat top human players in Go within a decade or two. Previous computer programs never played at a top level in this game.

To understand the significance of AlphaGo's achievement, one should look at how computers did in the game of Chess. In 1997, IBM's Deep Blue beat the then world champion Garry Kasparov 3½–2½ in a six-game match. Deep Blue was equipped with basic knowledge of Chess, such as "a Queen is more valuable than a Rook", and "controlling the centre of the board is more important than controlling the edges".

The basic approach in Deep Blue was searching: it evaluated the quality of each sequence of moves from the current board position. Searching each sequence to the end of the game is out of the question, due to the astronomical number of sequences available. Deep Blue used an intelligent algorithm to determine when to stop exploring a sequence. To evaluate the quality of a sequence, it used a hard-wired function to evaluate how favourable the board situation is (while hard-wired, this function was changed between games against Kasparov). It also used a clever search method to save computation time – by discarding provably inferior moves. But the key to Deep Blue's success was its computation power. With multi-processors and specialized hardware, it managed to evaluate 200 million board situations per second. This allowed Deep Blue to look six to eight moves ahead in normal situations, but over 20 moves in critical situations. A lot of human expertise in the game was deployed in Deep Blue.

The number of possible sequences in a Go game dwarfs that in a Chess game. Following White's 20 possible moves in its opening move, Black has 20 moves to choose from. Depending on the moves made so far, the number of possible moves next is roughly in the order of 30. A human player may, through experience, intuitively discard many of these moves, but it is hard for a computer program to do the same. So a Chess program treats all moves the same, except when it gains concrete evidence that a move is inferior. A Chess game seldom ends after 300 moves. Therefore, the number of possible sequences in a Chess game is nowhere near the possible sequences in a game of Go. If a Go program were to search, say, six moves ahead in every move, it would take too much time to compete in a match. For that reason, the intelligent search algorithm used in Chess will not go far in the game of Go. Besides, evaluating how favourable a board situation is in Go is harder than in Chess. Even top Go players often disagree on whether a board situation is favourable or not. Given the complexity of the game, even top Go players sometimes make moves that they "feel" right – they do so more often than in Chess. In summary, the size of the problem plus the difficulties in assessing a board situation together makes Go a much harder game than Chess for computers.

AlphaGo learns to play the game by playing. It accumulates its experience in the form of weights in its artificial neural networks (to be elaborated in Section 3.2). Instead of examining move sequences systematically, as Deep Blue did in Chess, AlphaGo used a method called Monte Carlo Tree Search. Basically, it tries out random move sequences in order to evaluate the quality of each immediate move. While the moves are picked randomly, they are not picked with equal probabilities. They are biased by how successful the positions are in AlphaGo's experience. That means a move that was found to be successful in the past in a board situation will be tried more often than the less successful moves. The more games AlphaGo plays, the more experience it gathers, and the better it plays.

2.2 GENERAL AI: THE ROSE GARDEN

Following the success of AlphaGo, DeepMind went on and developed a new version of the program called AlphaGo Zero. By dropping all the human input to AlphaGo, Zero started with the status of 19×19 positions (whether they are empty or occupied by black or white) to the artificial neural network. In other words, AlphaGo Zero learned everything from scratch: no opening books, no initial knowledge of favourable shapes, and no knowledge of moves made by experts in previous games. It learned through nothing but playing. Naturally, it played badly at the beginning. But it improved quickly. Eventually, it played a better game than AlphaGo. By beating AlphaGo 100–0, AlphaGo Zero established itself as the world's top player in Go. No human player has ever been able to beat AlphaGo Zero. This is a remarkable achievement.

Machine learning requires a lot of computation. The version of AlphaGo that beat Lee Sidol used multiple processors and GPUs.[2] Neural network based machine learning involves even more specific calculations than image processing which GPUs are built for. Encouraged by the success of AlphaGo, Google built specialized GPUs called Tensor Processing Units (TPUs) to speed up machine learning.

The success of AlphaGo Zero sparked Google's interest in "General Artificial Intelligence". The observation is that human input has always been the bottleneck in software development: the engineers have to program into the system knowledge about how to do the job well (which is referred to as "domain knowledge" – knowledge about the domain that the program is applied to). This is time-consuming and expertise-demanding. The hope is that machines can be made to learn everything from scratch without human input, as AlphaGo Zero did.

If General Artificial Intelligence succeeds, machines can learn everything by themselves. They can learn independently, through observation and interaction with the world. For example, by observing enough diagnoses, prescriptions and patient responses, machines can learn to be good doctors, perhaps better than human doctors. That is the vision of Google and its followers.

2.3 COMPLICATION: THE REALITY

AlphaGo Zero was a great success: it learned with the bare minimal amount of information about the game. It is therefore understandable why Google wants to extend this technology to make machines learn everything. The question is: how difficult is it to do so?

To answer this question, one has to understand that in the board-game Go, a board situation is fully described by the state of each position on the 19 × 19 grid: whether it is empty or occupied by black or white. Therefore, the minimal amount of input to AlphaGo Zero is the state of these 19 × 19 positions. Machine learning people refer to these inputs as "features", "attributes" or "variables". In this book, we shall call them variables (a term in computer science). A variable may take different values. In Go, the value of a variable that represents a position could take one of three values: "empty", "black" or "white".

As explained above, in Go, we know exactly all the inputs that are relevant to the game. Unfortunately, this is not necessarily the case in every application. Sometimes, it is unclear what variables are relevant. Suppose we want to forecast the FTSE 100 Index tomorrow. What variables should we consider? Would the FTSE prices in the previous 10 years be sufficient? Should we take the daily closing prices? Or should we take the tick-to-tick prices?[3] Should we take the prices of the individual shares in the Index as well? What about the inflation rate, interest rates, pound to US dollar exchange rates and unemployment rates? Where do we stop? If we include all the variables available, even the cleverest machine learning system will take a long time to learn. This is because any variable could potentially interact with any other variable. Any of these interactions could be reflected in the price of the FTSE 100.

Sometimes, new variables can be created. For example, in finance, we may collect information about a company, such as its debt–equity ratio and price–earning ratio. In the health sector, we may take readings in passing the patient through additional tests. In other words, we may create variables when needed. So, unlike the boardgame Go, it is not always obvious what variables to take as input to machine learning.

Should one use as many variables as available? Probably not. In the above examples, collecting financial information is not cost-free. Putting patients through additional tests (in order to collect data) could cost money. Tests may cause inconvenience or suffering to the patient. Besides, using more variables tends to (though not always) demand more time in machine learning. If one could identify the most relevant variables, one stands a better chance of finding patterns quicker than one's competitors in finance.

In the game Go, win or loss is clearly defined. This may not be the case in general. In some situations, we may not know exactly what we want. For example, finance is about risk and return. Normally, high-return projects involve high risks. Which of the following projects would you pick?

Project 1: It gives a return of 50%, but there is a 5% chance of losing 50% of your capital.

Project 2: It gives a return of 10%, but there is a 5% chance of losing 10% of your capital.

Project 1 gives a higher return but a higher risk than Project 2. The choice depends on your risk appetite and constraints. People's appetite for risk is complicated.[4] The trade-off between return and risk is not linear. Specifying the goal of machine learning may not be a trivial task. This is a non-trivial issue, which we shall look into in Section 5.4.

There is one more important point to remember when we extend AlphaGo Zero's machine learning experience to other domains: Go is a two-player game. Machine learning relies on feedback. For feedback, AlphaGo Zero can assess its performance by playing against, say, AlphaGo Zero itself. Getting feedback through self-playing is relatively simple. This is not necessarily the case in other applications. If one were to learn a trading strategy, one must consider many players: the stock exchange, the regulators (which may change the rules), the central banks (which may change interest rates) and the competitors (which may change their strategies in response to the market). Most importantly, the share price of a company that the strategy is used to trade on may fluctuate due to news about the company. Therefore, the performance of the trading strategy may not directly reflect the strategy's decisions alone. In other words, even if the strategy is brilliant, it may still lose money due to other factors. In computing terms, such feedback is described as being "noisy". It is a lot harder to assess the performance of a strategy with noisy feedback.

2.4 COMBINATORIAL EXPLOSION, THE CURSE OF COMPUTATION

It is important to realize that there is no magic in machine learning. All it does is to find the relationships between the input variables and the outputs. In the case of Go, the inputs are the state of each position and the outputs are the promise of each position on the board. In learning a trading strategy, the outputs could be "buy", "sell" or "hold"; a more complex strategy may involve cash flow or risk management.

If the relationship between the input and output is simple, then it is easy to learn. The Momentum Trading Strategies mentioned in the previous chapter are one such example. Suppose whenever the 7-day moving average crosses the 21-day moving average from below, the prices will always continue to rise. In that case, given historical data, all the machine learning system needs to learn is that calculating the 7- and 21-day moving averages is useful. Then it will have to learn, from historical data, how to compare the two moving averages, which is not difficult.

Unfortunately, simple patterns disappear quickly as soon as traders apply them in trading. Competition drives traders to find complicated patterns. In complicated patterns, the machine learning system must use more variables and learn more complex relationships. For example, it may have to consider 2-day, 3-day, ..., 100-day moving averages too. Comparing 7-day and 21-day moving averages may be good for deciding when to buy, but for selling, it may be better to consider, say, 10-day and 20-day moving averages. On top of that, in trading the shares of a company, it may be useful to consider the momentum in the index as well.

New intermediate variables could be created to explicitly represent the relationship of the primary variables. By considering more intermediate variables and more relationships between them, the number of possible combinations between them increases rapidly. This rapid increase is called the "combinatorial explosion problem". This is a fundamental problem in computation.

One way to understand the impact of the combinatorial explosion problem is to look at passwords. Passwords are useful because it will take a long time if one attempts to break them by trial and error. Here is an analysis: suppose a password is made up of eight characters, which could be numbers or letters or symbols. It could be one of the over 700 trillion possible combinations.[5] Even if one is able to try one million combinations per second,[6] one would need roughly 23 years to try all the combinations. If the number of characters is increased from eight to nine, then it will take 1,600 years to try all the combinations under the same assumptions. Increasing the length of the password to ten characters would increase the trial-and-error time to 43 million years. The number of combinations increases exponentially as the length of the password increases.

The combinatorial explosion problem is a fundamental problem in computation. The amount of computation required grows exponentially as the size of the problem grows. Even the fastest computers today will not be able to exhaustively search for all possible solutions. In those situations, finding optimal solutions is out of the question. That is where algorithms matter. A clever algorithm may stand a better chance to find better solutions than a poor algorithm. Alternatively, a clever algorithm may be able to find good solutions in a shorter time. AlphaGo explained above is a good example of a program that manages to find better moves than its opponents within the time constraints in the game of Go.

Earlier we explained that the main competition in trading is to find opportunities ahead of others. Due to combinatorial explosion, exhaustively searching all possible relationships among the variables is out of the question for any patterns of reasonable complexity. To find opportunities ahead of one's competitors, one has to deploy good algorithms and knowledge of how to deploy them. That depends on the quality of the machine learning expertise. Machine learning works with variables. Knowing what variables are relevant is also important. That depends on the quality of the financial expertise. So, as explained in the previous chapter, expertise in both finance and computing is important in this competition. That is all due to combinatorial explosion.

Before we end this section, it is worth asking: does the speed of hardware matter? Does it help to use multiple processors? Would they help to contain combinatorial explosions? The quick answer is: no, they cannot contain combinatorial explosion, but yes, they could still be useful. Here is why: the number of possible sequences in Go is bigger than the number of molecules in the universe. Increasing the speed of a computer by 1,000 times does not make the situation better. Having said that, if two traders use exactly the same algorithm (which is most unlikely to be the case at a professional level), then speed matters. More importantly, instead of taking 10 hours to learn a pattern, a trader would be very happy if it can take 10 minutes to do so. For that reason, faster hardware or multi-processors help. That is why it makes sense for Google to build TPUs (as mentioned in Section 2.2).

2.5 A MISSING INGREDIENT IN CLASSICAL ECONOMICS

To fully understand how computation could help, it is important to understand the role of computation in classical economics. Following are some of the most important assumptions in classical economics and their implications from a computation point of view:

- The perfect rationality assumption: everyone will make the best decision to maximize their interest.
- The homogeneity assumption: everyone will make the same decision at the same time.
- The perfect information assumption: everyone has access to all information in the market.

Economists all understand that these are simplifying assumptions. They know very well that no one is perfectly rational, traders and investors are not homogeneous, and information does not flow freely. They believe that these assumptions approximate reality.

All major economic theories are based on the above assumptions. The consensus is that if these assumptions are close to reality, the theories constructed under them are good enough to reflect reality. It is believed that the relaxation of these assumptions should not change the established theories too much.

Starting with simplifying assumptions is a common practice in science. This allows researchers to focus on the key issues and gain a good understanding of the subject. When the study is mature, scientists will relax the assumptions bit by bit to see how the established theories should be modified.

Unfortunately, the above assumptions are not exactly close approximations to reality. They ignore the importance of an ingredient: computation. This will be clear if we look at their computational implications:

- The perfect rationality assumption:
  From a computational point of view, if the perfect rationality assumption holds, everyone will make the optimal decision in every problem with regard to what it knows. As explained in Section 2.4, due to combinatorial explosion, many problems cannot be solved to optimality within one's lifetime. For example, no one knows what the optimal moves are in a game of Go, although this game involves no uncertainty or hidden information. How realistic is the perfect rationality assumption? Strong though AlphaGo is, it is unlikely that it plays the perfect game (if it does, it could not have been able to continue to improve). As will be explained later in this book, financial problems are much harder than a game of Go. Besides, most financial problems are time constrained (for example, the US dollar to Euro exchange rate changes rapidly). Solving problems to optimality in every problem within the time

constraints is out of the question. Even finding near-optimal solutions is a big ask.

- The homogeneity assumption:
We know that some algorithms are more efficient than others. That is why a whole body of research in computer science emerged: to study what classes of problems are tractable and what are not and how to search for solutions efficiently. For optimization problems which cannot be solved to optimality, one normally settles for the best solutions that one could find; some algorithms and heuristics would find better solutions than others. Even if two algorithms find solutions of the same quality, one may be 1,000 times faster than the other. Not all algorithms and heuristics are known to everyone. Specialized algorithms and heuristics have been designed to solve specific problems. Specialized algorithms and heuristics take time and expertise to develop. Homogeneity in problem-solving is far from being true.

- The perfect information assumption:
Not everyone has access to the same data at the same speed. Even if data is available freely, data scientists know that information costs. Expertise is required to extract information from data. Some will be able to extract more information from data than others. Computation power matters too. That is why Google builds specialized hardware for machine learning. It is unlikely that anyone would be able to acquire "perfect information". To gain perfect information, one has to maintain "consequential closure". This means if one knows that "A is true" and "A implies B", then one must infer that "B is true". If one also knows that "B implies C", then one must also infer that "C is true". Maintaining consequential closure means making all possible inferences; in other words, explicitly stating everything that one knows based on what one already knows. Nobody does that. Why? That is again due to

combinatorial explosion – there are just too many inferences to make. Anyone who maintains consequential closure should not hold conflicting beliefs. Most of us do hold conflicting beliefs (which we discover from time to time). For all these reasons, most of us only extract shallow information from data. Perfect information is beyond our reach.

All the above classical economics assumptions take computation for granted. They ignore the impact of computation. If how optimal a decision is defined by how rational one is, then one could say that the algorithms and heuristics that one uses determine one's rationality.[7] Different computer scientists know different algorithms and heuristics. Therefore, naturally, the homogeneity assumption does not hold. Besides, the perfect information assumption cannot hold as some algorithms and heuristics will extract more information from data than others. From a computational point of view, all the above assumptions are pretty far away from reality.

2.6 NEITHER CAN LIVE WHILE THE OTHER SURVIVES

Computer scientists study what types of problems can be solved efficiently and what are intractable by nature. They also study algorithms and heuristics that may solve certain problems faster and find solutions closer to optimality. If the classical economics assumptions above hold, then computer scientists' research in complexity and algorithms is irrelevant.

Machine learning dominates today's research in AI. All machine learning involves searching in a huge space of solutions – moves in the case of AlphaGo and patterns in the case of forecasting. If the perfect rationality assumption holds, they should find optimal solutions. We know they cannot in most problems. If the homogeneity assumption holds, then all programs should find solutions of the same quality at the same speed. AI researchers know that this is not true.

If the perfect rationality assumption holds, then much of the computer science syllabus could be scrapped. This includes complexity theory and algorithms and a substantial part of AI, including machine learning. Research in quantum computing is probably irrelevant too, as one of the motivations for developing quantum computing is to contain the combinatorial explosion problem.[8]

On the other hand, if the above assumptions do not hold, then most classical economics theories must be rewritten. If we relax the perfect rationality assumption, we need a quantitative definition of human rationality. Unfortunately, we do not have such a definition. Without a definition of human rationality, we do not know how to revise economic theories with the homogeneity assumption relaxed. To relax the perfect information assumption, we need to cost information, including information that we are yet to acquire. That means if we relax the above assumptions, we do not know how to revise classical economic theories.

So, should we continue to make the above assumptions, knowing that, from a computational point of view, they do not remotely describe reality? Or should we thoroughly revise classical economics given that they are built on shaky grounds?[9]

2.7 SUMMARY: POWERFUL BUT NOT MAGICAL

AlphaGo was a great success. AlphaGo Zero raised public expectations toward general AI, in which machines could learn by themselves with minimal human input. A rose garden was painted, which is good for AI funding. However, scientists must pay more attention to both promises and difficulties.

In reality, extending AlphaGo Zero's experience to learning everything is non-trivial. Unlike Go, the variables in real-life problems may not be obvious. Cost may be involved in creating useful variables for general machine learning. For example, in the health sector, new tests may be needed for prognosis; such tests may be expensive and unpleasant. In finance, extracting information from data costs, as explained in Section 2.5. Besides, feedback, which is the key issue in machine learning, may be noisy in learning trading strategies, as explained in Section 2.3.

One of the important take-home messages from this chapter is that while computers are fast, combinatorial explosion prevents them from solving many problems to optimality. That is where algorithms matter. Cleverer algorithms tend to find better solutions quicker.

In machine learning, the quality of data matters. That is where financial experts may help by providing useful variables. Different machine learning methods work with different types of data and different tasks. To apply machine learning to finance, the best way to succeed is to use a team with expertise in both algorithms and finance.

Classical economics is built on important assumptions: perfect rationality, homogeneity and perfect information. Economists know that they are simplifying assumptions. But they may not realize how remote these assumptions are to reality. These assumptions miss one important ingredient: computation. From a computational point of view, given combinatorial explosion, these assumptions are all unrealistic. If they were to hold, a large part of computer science, including AI, would become irrelevant. This is a complex issue, which will be revisited in the rest of this book. We close this chapter with the following statement:

> *Between classical economics and AI, neither can live while the other survives!*[10]

NOTES

1. 381 factorial means 381×380×379× ... ×3×2×1, which is an astronomically large number.
2. Graphics Processing Units (GPUs) are computer hardware specialized in processing images. Displaying 3D images onto a screen requires a lot of

matrix and vector calculations. Designers exploit the characteristics of their special operations to make them much faster than general processors.

3. The tick-to-tick prices means the traded prices in every transaction.

4. The size of the investment matters. Buying a random stock in the market is probably a better investment than buying a lottery. But many would buy a lottery because the money outlay is relatively trivial.

5. Suppose a password is made up of eight characters, which could be 0, 1, 2, ..., 9, A, B, ..., Z, a, b, ..., z or one of ten symbols (such as /, *, & or %). There are (10+26+26+10=) 72 possible choices for each character. The number of combinations is therefore 72×72× ... ×72 (8 times), which is over 722 trillion.

6. This is a generous assumption. Systems typically delay retries, so there are normally far fewer tries per second.

7. Herbert Simon (who won his Nobel Prize in Economics in 1978) acknowledged that human beings have limited rationality, which he called "bounded rationality". However, there has been no consensus on its definition. Perhaps the algorithms and heuristics that one uses define one's bounded rationality.

8. One could argue the opposite: when quantum computing is ready, the combinatorial explosion problem is contained. Then the perfect rationality assumption is closer to reality. Therefore, perfect rationality is work in progress. We shall have to see how quantum computing matures before we can conclude this complex analysis.

9. Behavioural finance and computational finance are two examples of research that relax the three classical economics assumptions.

10. The phrase "Neither can live while the other survives" is borrowed from J.K. Rowling's *Harry Potter and the Order of Phoenix*, 2003.

3

MACHINE LEARNING IN FINANCE

3.1 MACHINE LEARNING FOR FORECASTING

So far, we have introduced the synergy between computing and finance in algorithmic trading. We have also explained the success and limitations of machine learning. In this section, we shall take a closer look at an application of machine learning in finance: forecasting.

Forecasting is an important subject in finance. The hope is to predict what is going to happen based on what has been observed so far. We shall focus on the following forecasting target, which we introduced in Chapter 1:

Forecast Target 1: "Will the FTSE 100 Index rise by 4% within the next 7 days?"

The first step in applying machine learning to forecasting is to define the target. In the above example, the target is to predict whether "the price will rise by 4% within the next 7 days". In technical terms, the target of this forecast is to predict the value of a Boolean variable, which could take the value "true" or "false".

Readers are reminded that the forecasting target does not have to be a Boolean variable. One could try to forecast:

Forecast Target 2: "What will the closing FTSE 100 Index be tomorrow?"

DOI: 10.1201/9781003348474-3

In this case, the variable to be forecasted is a number.

After deciding what to forecast, the second step is to identify the variables that may help us to forecast the target. For example, technical analysts believe that future prices can be predicted from the past. In order to predict the value of the FTSE 100 Index tomorrow, they may feed a machine learning system with, say, the past 10 years' Index values. Fundamental analysts will use fundamental information, such as interest rates, exchange rates, consumer price index, trade surplus or deficit, etc., for forecasting.

Another decision to make is to decide on what machine learning method to use to predict the target with the input variables. Artificial neural networks were used in AlphaGo. Both artificial neural networks and genetic programming have been used in financial forecasting. Apart from these two, many other machine learning methods have been invented. Different techniques are suitable for different applications, so knowledge in which method works for what problems is important. This is a non-trivial topic and ongoing research, which is way beyond the scope of this book.

In the next section, we shall explain how machines might learn to predict the target values. Computer scientists call the form of learning that we are about to introduce "supervised learning". This is because it requires a training session, in which the trainer must tell the system what the correct target value should be for each set of input data. Later in this chapter, we shall introduce "unsupervised learning", in which no training is required.

3.2 SUPERVISED LEARNING

A few technical terms should help understand what machine learning is about: The input variables are called "independent variables". The target is called a "dependent variable", as for machine learning to work, the target's value must be dependent on the value of the input variables. If it is not, then machine learning will never find anything useful. Machine learning's task is to find the dependency relationship. That means finding out what the target's value should be given a set of input values.

In supervised learning, the trainer tells the machine learning system what the correct target value should be under a given set of input values. How would the trainer know what the correct output should be? That is normally done through hindsight. By looking back 7 days, one knows "whether the price will rise by 4% within the next 7 days" for "Forecast Target 1" above. For example, by looking back 1,006 days, one gets 1,000 sets of correct input–output relationships.

When the trainer provides the machine learning system with an input and tells it what the correct output should be, the system will adjust its internal parameters to guide it towards giving the correct answer in the future. That is how learning progress is made. Normally, there are many ways to adjust the internal parameters towards the given answer. There is no guarantee that the system will make the right adjustments. That is why the system will guide itself towards the correct answer rather than make sure that it gives the correct answer. This gradual learning strategy is used by most learning systems. It is also needed to allow the data to be noisy (sometimes the "correct answer" may not be correct in other cases; it could be an exception), which financial data often are.

The training is repeated with the same data. Every time the training data is passed through the machine learning system, it adjusted its parameters a bit. When training is completed, the output is expected to match most of the correct answers. This should be the case if the value of the target is indeed dependent on the input values.

Mathematically minded readers may see supervised learning as a "function-fitting" exercise. Learning is conducted through calibration. The machine learning system attempts to find a function that maps the input to the output. This function could take any form. In an artificial neural network, the function takes the form of a mathematical relationship between the input and the output via some intermediate internal variables. The neural networks that AlphaGo uses take many layers of internal variables, as shown in Figure 3.1, hence the term "deep learning". The input nodes (on the left of Figure 3.1) pass their values to the nodes in the first hidden layer next to it via the connections. Each connection carries a "weight", which adjusts the strength of one node to another. These weights are adjusted through training. The value of an internal node is the weighted sum of all its inputs. So, the whole network is a mathematical function from the input to the output.

With "genetic programming", another paradigm of machine learning, logical relations could be handled more easily. That means it can learn rules in the form of: "if the 7-day Moving Average was below the 21-day Moving Average yesterday, but the relationship is reversed today, then the price will rise tomorrow, otherwise …".[1] So, genetic programming can handle logical functions from the input to the output.

Anyone who uses supervised learning for financial forecasting must remember that they are making an important assumption: the behaviour of the market in the future is similar to the market's behaviour in the past. This is because training is conducted with historical data. If market behaviour changes, there is no guarantee that the forecasting function learned will apply to the market in the future.

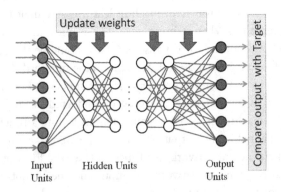

Figure 3.1 Structure of a Multi-layer Artificial Neural Network.

When would the market change its behaviour? New financial instruments play an important part: when futures and options were introduced, they gave traders more tools to conduct their trades. Technology plays an important part too. Algorithmic trading increases trading frequency dramatically. Political events change the traders' behaviour too. For example, when Brexit results were announced, the foreign exchange market temporarily behaved differently from before. These were the moments when the machine-learned forecasting may not have worked.

3.3 KNOW YOUR DATA

Machine learning algorithms are important, but supervised learning will not work unless it has data – not just in quantity, but in quality.

Big data matters: in general, the more data machine learning uses, the better the result tends to be. Supervised learning is basically a generalization exercise. It searches for patterns that are supported by data, with the hope that these patterns repeat themselves in the future. Generalizing from ten examples is dangerous. Generalizing from 10,000 examples is better. Everything being equal, the more data one uses, the more reliable the learned patterns.

However, "everything being equal" must be examined carefully. Suppose one uses daily closing prices for forecasting (discussed in Section 3.1). Using historical data for 750 days (about three years) for training is probably better than using 250 days. But is it a good idea to use 10,000 days (about 40 years)? Probably not. This is because the market from the financial crisis in 2007 and 2008 was very different from the market in more recent years. Patterns learned from that period may never repeat themselves in the future. Therefore, the inclusion of 2007–2008 data may not benefit machine learning.

Besides, the market has changed significantly. As mentioned above, algorithmic trading has grown in popularity; they trade differently from human traders. Besides, more financial instruments have been introduced. Regulations have changed, which affects the behaviour of corporate investors. All these make old data less relevant to today's market. Therefore, more data is not always better.

While we are on this topic, it is probably worth explaining why learning from "higher-frequency data" (such as minutely prices) is better than using "lower-frequency data" (such as daily closing prices). If one uses daily closing data, there are approximately 250 data points per year. If one uses minutely closing data, one has access to approximately 120,000 data points per year, nearly 500 times more. High-frequency finance will be elaborated in Section 7.2.

While having sufficient data is essential, the quality of data is critical to machine learning. In Section 1.5, we mentioned the lipstick index. Is it possible to use lipstick sales to forecast GDP growth in the next quarter? Correlations between lipstick sales and GDP growth may have been found in the past, but they do not hold all the time. To be able to forecast reliably, the variables used for prediction must be relevant to the target – the higher the relevance, the better.

Machine learning is no magic. How one prepares the data affects machine learning's ability to find patterns. For argument's sake, suppose the 7-day moving average is an important indicator which helps to forecast. If we input to the program the past 1,000 days' closing prices, machine learning could learn to use the 7-day moving average by itself. But if we pre-compute the 7-day moving average and supply it as an input variable to the program, then we increase the machine's ability to learn useful patterns using this variable.

To summarize, the data that one uses affects a machine learning program's effectiveness and efficiency. What a program may potentially learn depends on how financial data are collected and how we present the data to the program. We shall revisit the financial data issue in Chapter 6.

An old saying in computer science is always worth remembering:

> *"Garbage in, Garbage out!"*

Readers should note that this point in no way contradicts with AlphaGo Zero's idea of general intelligence (Section 2.2). The creator of AlphaGo Zero fed no intelligence about the game into the program. But AlphaGo Zero received all the input that is relevant to the game (the current state of each position). So AlphaGo Zero did not start with garbage; it started with all the necessary input for learning.

3.4 A GLIMPSE OF GAME THEORY

In this and the next section, we shall turn our attention to unsupervised learning. We shall use bargaining theory as an example to demonstrate how unsupervised learning works.

In supervised learning, the trainer must tell the program what the "correct" solution should be. For example, in the board game Go, AlphaGo used supervised learning to start. It used games played by top human players to show the program where good moves are. This allowed AlphaGo to conduct supervised learning at its initial stage. Following the success of AlphaGo, AlphaGo Zero was developed. There, supervised learning was dropped. This makes sense, as AlphaGo was already beating top human players. Even if supervised learning were to be conducted, AlphaGo Zero should have been shown AlphaGo's moves, not human players' moves.

Unsupervised learning is conducted when we cannot tell (or, in AlphaGo Zero's case, do not want to tell the program) what the target solution is. It does not matter whether we know what a good solution is, as long as we know whether a solution is good or bad when we see it. In the game of Go, we know a good program is one that wins more games than it loses. This is sufficient for unsupervised learning to apply. In this section, we shall introduce a bargaining problem. In the next section, we shall explain how unsupervised learning can be applied to the bargaining problem.

Bargaining is one of the two most studied areas in game theory (the other being repeated games, such as the Prisoner's Dilemma). In the next section, we shall look at how unsupervised learning could be applied to bargaining. Before that, we shall introduce a bargaining model.

Following is a textbook scenario in bargaining:

Basic Alternating-Offers Bargaining Model

Players A and B are about to share a pie. A makes an initial offer, offering to share a certain percentage with B. B may reject the offer, in which case, B will make a counteroffer to A. Then it is A's turn to decide whether to accept or reject the offer. The bargain continues until one side accepts the offer or no deal is struck. To give both parties an incentive to make reasonable offers and accept offers as soon as possible, the game stipulates that the utility of their shares drops increasingly over time. Importantly, both parties know how fast both utilities drop. The player whose utility drops more slowly would have an advantage in the bargaining.

Here is a bargaining scenario:

Turn 0: A offers 20% of the pie to B.
  If B accepts the offer, then A will have a utility of 80% and B 20%.
Turn 1: B rejects A's initial offer; B counteroffers 50% to A.
  If A accepts the offer, then A will have a utility of 27% (not 50%, as the utility drops) and B will have a utility of 34% (assuming that B's utility drops more slowly)
Turn 2: A rejects B's 50% offer; A counteroffers 40% to B.
  If B accepts the offer, then B will have a utility of 18% (discounted from the 40% being offered) and A 18% (discounted from the 60% that A retains)

  . . .

A little reflection should convince the readers that A would have been irrational to reject B's offer of 50% in Turn 1 if A planned to counteroffer 40% to B in Turn 2. This is because accepting the 50% in Turn 1 gives A a utility of 27%, which is better than getting 18%

(discounted from 60%) in Turn 2. A would have been better off taking the 50% offer from B in Turn 1.

To push the logic further, A should have made B a better initial offer (in Turn 0) had A anticipated that B will reject 20%. In order to mathematically work out what A's initial offer should be, bargaining theorists assume the following:

Assumption 1: Both players are fully rational, which means they can both make decisions that maximize their share of the pie (see discussion in Section 2.5). Both players know that their opponent is fully rational.

Assumption 2: Both players know the rates at which the two players' utilities drop over time (their utilities may drop at different rates). They also know what their opponent knows.

Under the above assumptions, bargaining theorists can work out A's initial offer that B cannot refuse. Both players will be better off agreeing upon the initial offer. Any delay in the agreement will discount their utilities. In other words, while the two players are competing for limited resources, they must also cooperate in order to maximize their rewards.

It is worth reminding computer scientist readers that the search for the initial offer is not a simple optimization problem. A computer scientist would be tempted to try one value at a time, from 0% to 100%, for a given precision in an attempt to find the optimal solution. However, it is not a simple problem of evaluating every value because how good an offer depends on the opponent's response, which in turn depends on the first player's subsequent response. Computationally, this is not dissimilar to a game of Chess or Go. Game theorists call the solution a "subgame equilibrium" (as opposed to an "optimal solution", which computer scientists are familiar with). To make the initial offer at Turn 0, Player A would ask itself what B would offer in Turn 1 if B rejects A's initial offer. In other words, Player A attempts to solve B's subproblem at Turn 1.

With that subgame solution, A would know what to offer in Turn 0. But then how would B solve the subproblem in Turn 1? A would anticipate that B would solve the subgame problem at Turn 2 from A's point of view. So the subgame equilibrium is solved recursively. When the reasoning is repeated recursively towards infinity, which is possible mathematically, the initial offer at Turn 0 can be solved.

3.5 "UNSUPERVISED LEARNING" FOR BARGAINING

In this section, we shall explain how unsupervised learning can be applied to the Basic Alternating-Offers Bargaining model introduced in the previous section. The approach that we are going to explain is based on "evolutionary computation", an idea that is borrowed from natural evolution. Given a problem, instead of designing and building solutions, one attempts to evolve solutions.

To apply evolutionary computation, candidate solutions must be represented using building blocks — think of them as genes. (Knowledge representation is an important part of AI. We shall revisit this issue in Chapter 6.) For this bargaining model, a candidate solution is a function made up of the two players' discount rates which determines how fast the two players' utility drops over rounds. Computer scientists are familiar with representing functions with trees. So a candidate solution can be represented by a tree which branches are made up of arithmetic operations (+, −, ×, ÷), numbers and the given discount rates. Different combinations of these operations form different functions. A tree is a function that can be reduced to a number that represents a player's offer to their opponent. The task of machine learning here is to explore different ways to combine the building blocks.

The general principle of evolution is to maintain a population of candidate solutions and let them evolve good solutions: To start, a population of candidate solutions is generated randomly. The fitness of the individuals in the population is determined by how well they meet the requirement (in the case of optimization) or solve the problem (in the case of problem-solving). The fitter an individual, the more chance it is given to pass its building blocks (genes) to future generations. The hope is that building blocks that contribute to fit individuals will be allowed to construct better solutions. Evolution ends when individuals in the population converge on similar solutions (this will be the case when the majority of the individuals use the same building blocks), or time runs out.

To apply evolutionary computation to the above bargaining model, a two-population approach can be used. Based on the solution representation described above, a population of strategies is generated for Player 1 and another population for Player 2, as shown in Figure 3.2. These two populations co-evolve through competition between the individuals. Individuals in the population for Player 1 will bargain with individuals in the population for Player 2. Successful individuals, namely, those that score high utilities will be encouraged to pass their building blocks to future generations.

This approach to learning bargaining strategies is a form of unsupervised learning. Unlike supervised learning, there is no trainer to tell the program what the correct solution should be. The candidate solutions find their fitness through playing against opponents. The designer's task is to design the representation of candidate solutions and the way to maintain evolutionary pressure to enable fit individuals to pass their building blocks (genes) to future generations.

It is worth iterating the point that if the programs are allowed to evolve solutions freely, many of the candidate solutions generated could be very poor bargainers. They could ask for over 100% of the

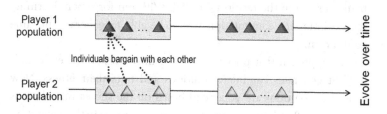

Figure 3.2 Co-evolution in Bargaining.

pie. They could also ask for a negative percentage of the pie. To help the search to focus, incentives or constraints can be added to the fitness evaluation step to help the search to focus on more promising solutions. Details of this approach are beyond the scope of this book.

There are at least three reasons why machine learning is an attractive approach to handling bargaining problems. Firstly, classical bargaining theory is a mathematical approach. It relies on the perfect rationality assumption. Human beings are not perfectly rational, as Nobel laureate Herbert Simon pointed out (discussed in Section 2.5). Human bargainers rarely can reason recursively towards infinity (as explained in the previous section). Instead of assuming perfect rationality, machine learning assumes reinforcement learning. This is arguably closer to human reasoning.

Secondly, a mathematical approach can only handle neatly defined mathematical problems. Unlike the Basic Alternating-Offers Bargaining model, real-life bargaining problem often involves messy relations, including logical and procedural operations, which makes mathematical analysis very difficult. A machine learning approach will handle logical and procedural operations all the same.

Thirdly, a slight alteration of the bargaining model could demand a completely new mathematical analysis. With evolutionary computation, one only needs to change the bargaining strategy representation. The evolutionary process is the same. For example, if Player B is ignorant about the utility deterioration rate of Player A, all one needs to do is remove A's utility deterioration rate from the language that defines Player B's strategy representation. With B's ignorance, a mathematical analysis would find the subgame equilibrium difficult to solve in the resulting model. With reinforcement learning, the evolutionary computation would be able to generate subgame equilibrium.

The approach that unsupervised learning uses is "generate-and-test". It generates candidate solutions and tests their fitness. New candidate solutions are generated based on the successful solutions found so far. Randomness almost always plays a part in the generation of new solutions. Randomness is important because the approach is basically sampling in the space of solutions. "Generate-and-test" is an old AI term. It was forgotten because it does not sound as exciting and imaginative as other terms such as artificial neural network, evolutionary computation and other search methods that use names which suggest nature inspiration. But it describes what many search methods, including unsupervised learning, basically do.

3.6 SUMMARY: MACHINE LEARNING IS A GAME CHANGER

In the previous chapter, we explained the promise and limitations of machine learning. In this chapter, we have looked into two forms of machine learning: "supervised learning" and "unsupervised learning".

Supervised learning requires the trainer to tell the program what the correct target values are. For example, in forecasting, the trainer must tell the program what the correct forecast is. What supervised learning does is essentially function-fitting: the machine learning system uses the training material to calibrate a function that would produce a forecast from the input variables. For supervised learning to be successful, it is crucial to choose the right variables: the value of the target must be dependent on the value of the input variables.

Unsupervised learning does not require a trainer. Solutions are evolved rather than designed. It has been used by AlphaGo to play the game of Go; it has also been used to find subgame equilibrium in bargaining, a branch of game theory. What unsupervised learning does is essentially generate-and-test: successful candidate solutions are encouraged to pass their building blocks to future generations, with the hope that better solutions will be evolved over generations. For unsupervised learning to be successful, it is important to build a proper (artificial) environment for the individuals to interact within and a reliable assessment of an individual's fitness.

Machine learning is powerful. It can be a game changer. However, one must understand that there is no magic in machine learning. Before the General AI approach (described in Section 2.2) succeeds, expertise is needed to help machine learning to succeed. Expertise is needed in determining what to learn, choosing the variables, designing candidate solutions, choosing machine learning methods or developing new ones if necessary.

NOTE

1. This is an example showing the form of a logical relationship between the input (which are 7-day moving average and 21-day moving average values) and the output (which is "will the price rise tomorrow?"). Readers are reminded that this is just an example, not a realistic rule.

4

MODELLING, SIMULATION
AND MACHINE LEARNING

4.1 MODELLING

A model is an abstract description of a subject. Here the subject
could be anything from a situation (such as a conflict), a system
(e.g. a banking system) to the dynamics of a market (e.g. an auction
market). To build a model of a subject means to identify the key
components of the subject and describe the relations between them.
The hope is to use the model to capture the main behaviour of the
subject.

In a model, the components often influence or interact with each
other. Such influence or interacting relations could be expressed in
any form. They could be expressed mathematically or procedurally,
for example, "if component A gets a signal from component B, then
A will send signals to components C and D".

Modelling enables one to reason about the subject. We encounter
models all the time. For example, at war, army officers put model
armies on a map to show their control and influence. This enables
them to evaluate moves and counter moves. A war game is a model
of real wars. The game SIMS is based on a model of how people
interact with each other.

The Basic Alternating-Offers Bargaining model introduced in
Section 3.4 is a simple model of bargaining. There the components

DOI: 10.1201/9781003348474-4

are the two players who take turns to make offers and counter-offers to each other. An offer is a percentage of the pie that the player proposes to take. Following are two examples of modelling applied to finance and economics.

Modelling in Interbank Payments

Models have been built to study interbank payment systems. It is a subject studied by central banks around the world, especially after the 2007–2008 financial crisis. When a customer of Bank A pays £1,000 to a customer of Bank B, Bank A must at some point pay £1,000 to Bank B. However, later in the day, another customer of Bank B could be paying a customer of Bank A £800. If the two banks clear their balances by the end of the day, all Bank A needs to do is to pay Bank B the difference, which is £200. The only drawback of doing so for Bank B is that if Bank A goes bankrupt during the day, B will lose £800 which it has already paid its customer. To avoid this risk, the two banks may clear the interbank payments instantaneously. The drawback is that they must maintain a reserve to do so, which eats into their profitability. In the above example, Bank A only needs to use a reserve of £200 to clear the payments by the end of the day, but it must use a reserve of £1,000 were it to clear the balance instantaneously. Central banks have gathered together to design, with the help of models, clearance rules to balance between risk-bearing and reserve burdens.

Modelling in Electricity Markets

Models have been built for designing the rules that govern an electricity market. The electricity market is complex. Multiple suppliers generate electricity to supply end-users through the grid, which distributes electricity to consumers. Excess electricity supply cannot be stored in large quantities and therefore goes to waste. However, a blackout is possible if electricity is under-supplied; the 2000–2001 California blackout was a well-known example. Government regulations and the rules of the electricity market must be designed to ensure a surplus in supply but minimize wastage (costs will eventually be passed on to consumers, so high wastage means high electricity prices). Designing the rules for the electricity market to strike a balance is non-trivial. This has been the subject of research in modelling.

4.2 MODELLING: IMPERFECT BUT USEFUL

Faced with a complicated situation, one often asks: "where should I pay attention to?" Model building helps people to identify the most relevant components and their relations in complicated situations.

In building a model, one is forced to ask what the key components are in the subject, and how these components relate to or interact with each other. For example, in modelling a war situation, the firepower and range of a troop may be quantified. If the modeller believes that the terrain is important, then objects such as rivers, grassland, trees and buildings should be part of the model. If the weather situation is considered to be important, then objects such as rain, snow, wind direction and wind speed should be part of the model too.

Model builders often start with the most basic components and relationships. They knowingly leave out less important components and their relations for a later stage. The initial models are naturally imperfect. A simple model is opted for because it is easier to study. After studying the simple model, more components can be added. More relationships between the components can be added too. An incremental approach enables the modeller to assess the impact of each additional component and relationship.

As more components and relations are added, the model is closer and closer to reality. However, most situations worth studying are complex, hence a model is never a perfect description of reality. We all know that the Basic Alternating-Offer Bargaining model does not describe human bargaining realistically. Communication in human bargaining is a lot more complicated. For example, in a market, a buyer may walk away, hoping that the seller will call him/her back with a better offer. Eye contact and body language are also important in human bargaining which is not in the basic bargaining model.

While a model is never a perfect description of reality, it can still be useful. Building a simple bargaining model helps one to focus on the factors that are most important and study the orders in a subject. When research in the simple bargaining model matures, bargaining theorists may incrementally relax the assumptions or refine the model to make it more realistic.

Due to complexity, one may never be able to build a very realistic model. But what is the alternative? An incremental approach is arguably the only way to study a complex situation. A model is always a simplification of the real situation. All models miss out on something. However, when used properly, models can be useful, as has been demonstrated in many applications.

"*All models are wrong, but some are useful*". (George Box)[1]

The topic of modelling will be revisited in Chapter 5 when we discuss portfolio optimization, a financial application.

4.3 SIMULATION: BEYOND MATHEMATICAL ANALYSIS

Models support analysis. When a model is simple, one may be able to mathematically analyse its properties. For the basic bargaining model described in Section 3.4, game theorists have been able to mathematically work out the subgame equilibrium under perfect rationality and perfect information assumptions.

Unfortunately, many interesting models in finance and economics are complicated. For example, in the bargaining model, what if one player does not know the other player's utility decreasing rate? Where would the subgame equilibrium be? That is difficult for mathematical analysis.

When the model is too complicated for mathematical analysis, simulation is often the only reasonable solution. Simulation is used by AlphaGo (explained in Section 2.1) – it runs through millions of moves by the two players in order to evaluate the quality of each possible next move. The unsupervised learning approach to the bargaining problem (explained in Section 3.5) is also a simulation – it runs through possible reasoning by the two players. This kind of simulation is sometimes referred to as Monte Carlo simulation, to reflect the randomness in the process.

Following is a simplified version of how AlphaGo uses simulation to make a move: given a board situation, AlphaGo will generate a random move (to be elaborated below) for the immediate move. Then it generates a subsequent move by the opponent, followed by a subsequent move by the current player, and so on. This simulation brings the game to the end (when all board positions are uncontested), which will tell AlphaGo which side wins. This simulation is repeated millions of times. The immediate move that leads to more wins will be adopted to be the next move.

To improve the efficiency of simulations, AlphaGo does not pick every empty position on the board with equal probability when it generates the next move. More promising positions are given more chances to be picked in the simulation. This is akin to human players spending more time examining the most promising move sequences. Machine learning is used to learn which positions on the board are more promising.

Monte Carlo simulation is not the only way to conduct a simulation. The co-evolution in finding the subgame equilibrium in bargaining, introduced in Section 3.5, can also be seen as a simulation. There one population is used to represent a set of strategies that Player A could adopt, and another population for player B. The two players continually evolve their portfolio of strategies in response to the other player's evolution. Readers are reminded that the main motivation for using co-evolution for the bargaining problem is to relax the perfect rationality assumption in finding subgame equilibrium.

4.4 CASE STUDY: RISK ANALYSIS

After the global financial crisis from 2007 to 2008, the international banking community gathered in Basel, Switzerland, to construct frameworks for securing financial stability in the future.[2] The Basel standards are implemented by individual countries in the form of laws and regulations for their banks. Banks were required to reserve a certain proportion of their capital to protect them to a certain limit in financial crises. The proportion depends on the assets that they hold – a lower reserve is required for assets of lower risks and a higher reserve for assets of higher risks.

The purpose of keeping the reserve is to reduce the chance of bankruptcy by the banks, which would disrupt society. Return on investment is not the regulator's concern. However, from a bank's point of view, the goal is to maximize its return. While it is the bank's duty to comply with the regulatory requirements, maintaining excessive reserves will eat into the bank's profit. In Chapter 5, we shall look closer at the problem of having dual conflicting objectives. In this section, we shall focus on how a bank may satisfy the reserve requirements. The description below is based on the implementation by a technologically advanced financial institute.

As reserves tie up capital, keeping excessive reserves reduces the banks' earning potential. For that reason, banks tend to carry the minimum amount of reserve to meet regulatory requirements. Banks are invited to present evidence to demonstrate that the reserves that they keep meet the requirements. This is where research is required.

Given a portfolio of assets held, a company will have to calculate the minimum reserve that it must hold. To do so, the company models the probability of each asset changing values. For example, if it holds a certain amount of bond X, a model may describe the probability of X losing 0.5% on the next day, the probability of X losing 0.4% the next day, ..., etc. How is this model built? It may be based on the historical price changes of X in the past, say, 3 years. Alternatively, the model may be built by using the statistical summary of X's price changes in the past 3 years – the mean and standard deviation, for example.[3] Alternatively, a mathematical model can be built based on the analyst's insight into that asset. For example, the analyst may expect the value of this asset to rise by 0.02% per day, with a standard deviation of 0.01%.

Apart from modelling the price changes in each asset that it holds, the company models the potential change in other external factors, such as the interest rate, inflation rate, etc. Again, changes in these factors can be modelled with historical data, mathematical data or expert insight. The company may also model the dependency between these factors and assets. For example, if the interest rate rises, share prices tend to fall. Such a relationship is nonlinear and complex, but that does not prevent machines from learning from historical data.

Having built these models, the company may start to simulate possible futures. Gains and losses over the next 30 days, say, can be simulated using those models. With these simulations, risk measures can be collected for the portfolio held by the company. For example, with over 100 million simulations, how bad could the portfolio perform in the worst 5% of the simulations? Suppose all these 5% of simulations show a loss of 8% or more, then −8% is called the 5% Value-at-Risk (VaR) of this portfolio, based on the models and simulations. The average loss of the worst 5% is called the 5% Expected Shortfall. These, plus other statistical measures from the simulations, can be presented to the regulators to show the company's reserve meets the regulative requirements.

4.5 ADDING MACHINE LEARNING TO MODELLING AND SIMULATION

We have explained in the preceding section that with models and simulations, a company can assess the risk of a portfolio. We have also explained the use of machine learning in AlphaGo to learn the promise of each position, which makes simulation more efficient. We have explained (in Section 3.5) how machine learning can be used to find the subgame equilibrium in a bargaining problem. In this section, we shall explain the power of adding machine learning to modelling and simulation. As an example, we shall describe how it can be used in designing trading strategies. Following is a cycle for designing trading strategies for a market.

A model–simulate–learn cycle for designing trading strategies:
1. Model the market clearing rules and other traders' behaviour.
2. Model a class of trading strategies, which is the subject of fine-tuning.
3. Simulate the interaction between the subject trading strategies and other traders in the market.
4. Assess the performance of the subject trading strategies.
5. Modify the trading strategies, guided by their observed performances.
6. Run the simulations again.
7. Repeat Steps 4–6 until the results match the desirable results.

To start, the investigator builds a model of the market clearing mechanism. This includes how orders are processed and how they are matched to complete transactions. Stock markets typically adopt a double queue system: the buyers form a queue and the sellers form another queue; traders transact with each other. Foreign exchange markets are typically market-making markets – the market maker sets the buying and selling prices, which they adjust based on supply and demand; traders transact with the market maker.

The modeller may also model trading strategies used by other traders. For example, some traders may use technical rules (they are called technical traders). Some may buy or sell randomly (they are called noise traders) or hold to take profit. As mentioned earlier, models are never perfect, but they could be useful for investigation.

A note on terminology: research that models players and their interactions in a system is sometimes referred to as "agent-based" research, which is a branch of AI. The word "agent" is used to refer to both human traders and algorithmic trading systems.

Next, the investigators must decide how to model a class of trading strategies which could be fine-tuned by machine learning (Step 2). For example, a trading strategy may take the form of a neural network with many internal layers (which is referred to as "deep learning" in AlphaGo, see Section 3.2). Machine learning will then be tasked to adjust the weights on the connections later in the process. The investigator may also decide to maintain a population of neural networks and let them compete against each other.

A trading strategy may also be represented by a tree. A tree is a generic data type which can be used to represent anything computable. Old-fashioned computer scientists were taught that all computer programs can be parsed into trees.[4] In genetic algorithms and genetic programming, which is a population-based machine learning method, a trading strategy is constructed by building blocks. For example, in genetic programming, a trading strategy could be represented by a tree, which is made up of arithmetic operations on financial indicators, such as the previous closing, opening, high and low prices. Trading strategies will be allowed to compete against each other. The poor performers will be eliminated and the fit individuals will be allowed to pass their building blocks on to future generations.

One important point in machine learning is worth reiterating: The choice of representation determines what one can learn. So is the choice of learning method. However, the most important decision is the choice of input and output. A technical trader may input to a learning system technical indicators, such as moving averages. An economist may input into the system fundamental indicators, such as price–earning ratios and macroeconomic indicators, such as interest rates. A poor choice in representation or machine learning method may still produce a mediocre trading strategy for the investigator. But, as explained in Section 3.3, a poor choice in the input variables will limit the ability of the system to produce anything useful.

Once the modelling is complete, the investigators may let the implemented trading strategies interact with each other (Step 3). The performance of each strategy is assessed (Step 4). The investigators must decide what they want to achieve in these traders. Performance could be measured in many ways. For example, it could be measured by the profits made, the percentage of profitable trades or the amount of maximum losses.

How does the system learn? How could it modify the trading strategies based on the feedback (Step 5)? With neural networks, the weights on the network connections record the accumulated knowledge acquired through learning. As in AlphaGo Zero (Section 2.2), this is a form of unsupervised learning. If a population of neural networks is maintained, then networks that performed well so far could be duplicated. Each copy will make minor random modifications to the weights of the connections. The new copies will replace the poor performance in the population. If a decision tree is used to represent trading strategies, a population of decision trees is maintained. Successful strategies are allowed to pass their building blocks to future generations, as explained earlier (see Section 3.5).

After the trading strategies are modified, the simulation will be repeated (Step 6). This simulation and remodelling process can be repeated until the investigator is satisfied with the strategies generated or no improvement is observed in the process.

Above, we have explained how the modelling–simulation–learning cycle could be used to automate the design of trading strategies. There are many other possible approaches. The key point is that the model modification step is laborious if the investigator were to be involved. Machine learning allows the system to test thousands of trading strategies, which would not be practical to do manually.

Some may ask: is efficiency that important? If a trading strategy works, then it is worth spending time to find it. This is a reasonable proposition. It is worthwhile to spend a whole year to find a winning strategy if this strategy reliably makes money in the market for the years to come. Unfortunately, a winning strategy will not be winning if others have found it too. Simple strategies such as the momentum rules and Head and Shoulder pattern probably worked at some point in the past. But as more traders know about them, these known patterns are reflected in the price. To succeed, an investigator must keep looking for new trading strategies. As explained earlier (Section 1.2), the competition is on finding regularities ahead of one's competitors. Therefore, efficiency in inventing new trading strategies matters. By automating the investigation process, the modelling–simulation–learning cycle helps to keep a trader ahead of the others in the game.

4.6 MECHANISM DESIGN

In the previous section, we explained how trading strategies can be designed through a modelling–simulation–machine learning cycle. In this section, we shall look at how rules can be designed for a new market. In economics, this is referred to as "mechanism design", a subject for the Nobel Prize in Economics in 2007.

New markets are created from time to time. We mentioned the electricity market in Section 4.1. This is a continuous market with changing supply and demand. The complexity makes its design challenging. In this market, the electricity demand over time is predicted based on demand patterns in the past. However, the weather, special activities (for example, when a football game is scheduled, electricity demand is expected to surge at half-time because many viewers will put the kettles on simultaneously) and other factors all affect the current demand. Electricity producers must bid the price they are willing to charge and the quantity of electricity that they are willing to produce. To avoid blackouts, the auctioneer (electricity supplier) must buy enough electricity to meet the demands continuously, with demands varying over time. On the other hand, the auctioneer wants the producers to bid the lowest prices under their individual business models. Different suppliers have different capacities. Small suppliers may carry higher production costs, but the auctioneer may not want them to be competed out of the market completely because their presence adds to the stability of the market. Suppliers that generate reusable energy may carry higher production costs too, but the supplier may want to support them for social reasons. In designing these rules, the auctioneer wants to give the producers enough incentive to produce enough electricity. They also want to encourage producers to bid their true valuation of the commodity.

The following is a modelling–simulation–learning cycle for designing rules in a market.

A model–simulate–learn cycle for mechanism design:

1. Model the market rules (which are the subjects of machine learning) and the participants' behaviour.
2. Simulate the interaction between the participants under the market rules.
3. Observe and analyze the results of the simulations.
4. Compare results with desirable results.
5. Modify the market rules in the market model, guided by the observed results.
6. Run the simulations again.
7. Repeat Steps 3–6 until the results match the desirable results.

This process is similar to the design of trading strategies, except in the focus on the models – here the focus is on the market rules (Steps 1 and 5).

Here, the model of the market is a generic framework which may have a set of rules to be selected or not selected, plus a set of parameters to be tuned (Step 1). Machine learning will be used to select those rules and tune the parameters (Step 5). The bidding and acceptance of bids are simulated (Step 2) with results observed and analysed (Step 3).

The auctioneer must decide what they want to achieve in the market. For example, how much blackout risk is it willing to take? That determines how big a surplus (buffer) it must maintain. How much does it want to support more expensive suppliers? That determines how it may accept bids. These objectives must be written down clearly for the modelling–simulation–learning cycle to be automated.

The objective function is used to evaluate the quality of a market model (Step 4). This is crucial to machine learning, which has to decide which models to modify and how to modify them (Step 5).

After the market models are modified, the simulation will be repeated (Step 6). This simulation and re-modelling process can be repeated until the investigator is satisfied with the market rules generated or no improvement (as judged by the objective function) is observed in the process.

What we have described earlier is one way to use a modelling–simulation–learning cycle to automate mechanism design in a market. Without automation, the model modification step could be laborious. Machine learning allows one to test thousands of market models, which would not be practical to do manually.

4.7 CONCLUSION: MODEL–SIMULATE–LEARN, A POWERFUL COMBINATION

Modelling helps one to focus on what to pay attention to. A model helps one to reason about a subject – such as a financial market or a payment system.

Mathematical reasoning is elegant and powerful. But it is only useful for simple situations – such as the Basic Alternating-Offers Bargaining model mentioned in Section 3.4. To study a complex system, simulation is cost-effective; sometimes, it is the only way. We have shown how modelling and simulation enable one to assess the risk of a portfolio.

Adding machine learning to modelling and simulation allows one to find subgame equilibrium in complex game models beyond the Basic Alternating-Offers Bargaining model.

To summarize: modelling, simulation and machine learning could combine to form a powerful tool. Modelling enables simulation. Machine learning helps to improve models. As models are never perfect, such reasoning is always flawed. However, having the means to reason about a subject is better than doing nothing at all.

> **"More calculation is better than less calculation; some calculation is better than none". (Sun Tze)[5]**

NOTES

1. G. Box and N. Draper, *Empirical Model-Building and Response Surfaces*, John Wiley & Sons, 1987.
2. Three Basel standards progressively covered more and more aspects of financial aspects. The latest standards were defined by Basel III in November 2010.
3. Provided that the price change distribution follows a Gaussian distribution or other statistical properties.
4. Parsing is no longer a popular part of the computer science syllabus today.
5. Sun Tzu, *The Art of War*, around 5BC ("多算勝，少算不勝，而況於無算乎？").

5

PORTFOLIO OPTIMIZATION

5.1 MAXIMIZING PROFIT, MINIMIZING RISK

A fund manager's task is to achieve two objectives: to maximize profit but at the same time minimize risk. In general, assets that give a higher return bear a higher risk. Government bonds are relatively safe because governments seldom go bankrupt (that happens occasionally). However, the interest that a government bond pays is generally lower than the return on stocks.

The best that a fund manager could do is to base their judgement on their knowledge about the assets available for investment. In other words, they can only do their best to find the expected return and risk for each asset. If their assessments are wrong, then they will make wrong decisions. How to forecast returns and risks is not the subject of this chapter. In this chapter, we shall examine what the fund managers could do based on their assessments.

Ideally, a fund manager should invest all their funds in an asset that pays a higher return and bears a lower risk among all assets in the market. Unfortunately, such an asset does not exist. Even if it does, it will disappear soon. This is because it will be snatched up by participants in the market in no time at all, which means the asset's price will rise, resulting in a lower return to future buyers. If such an asset emerges, and a fund manager spots it and acts

DOI: 10.1201/9781003348474-5

ahead of everyone else, then indeed that is what the fund manager should do.

In general, the fund manager will have to choose among a set of assets of varying returns and risks (according to the fund manager's assessment). They will have to decide how to allocate their funds to individual assets. This is referred to as the "portfolio optimization problem".

When the prices of individual assets in a portfolio change, the fund manager may find that the current portfolio is no longer the best according to their portfolio assessment criteria. The problem of how to adjust the current portfolio is referred to as the "portfolio management problem". To simplify our discussion in this book, we shall not go into this problem (we shall come back to it briefly in Section 5.6). Neither shall we discuss short selling, despite it being important in hedge fund management.

The common saying: "don't put all the eggs in one basket" applies to investment. Diversification is one of the basic principles in portfolio optimization. Dividing one's investments into holdings of two different shares could reduce risk, as long as the two company's share prices do not move up and down at exactly the same rate. If the share prices of the two companies always move in opposite directions, "market risk" (the risk resulted in the price movements in the market) is eliminated. In reality, share prices do not always move in the same or opposite directions. Diversification normally reduces market risk.

5.2 THE MARKOWITZ MODEL FOR PORTFOLIO OPTIMIZATION

The best-known approach to portfolio optimization is the Markowitz model. This model assumes that the fund manager is given a fixed set of available assets to invest in. The fund manager may invest any amount of its capital into any asset. No short selling is allowed. The goal is to maximize return and minimize risk.

Before deciding on how to allocate the funds, the fund manager must assess each asset's expected return and expected risk. The return of an asset may be based on its past return over a given period of time. There are many ways to quantify risk. In the Markowitz model, the risk is measured by the standard deviation of the returns over a period of time. Log returns are commonly used in practice. For daily data, the log return on each day will be calculated.[1] Then the standard deviation of daily log returns will be taken as the asset's risk over that period.

After establishing the return and risk of each asset, one must establish for every pair of assets whether their prices change together in the same direction, in opposite directions or independent of each other. This is measured statistically by the "correlation coefficient". The basic idea is that the more the two assets' movements agree with each other, the larger their correlation coefficient. That means statistically when one asset's price falls, the other's price is likely to fall as well. Holding both assets at the same time is riskier than holding two assets which correlation coefficient has a negative value, which means statistically their prices tend to move in opposite directions.

A portfolio is an allocation of funds into the assets available. For example, for a portfolio with three assets, the allocated funds to them, referred to as "weights", could be 25%, 35% and 40%, respectively. The return of a portfolio is the weighted average of the returns of the assets selected. The risk of a portfolio reflects how much is allocated to each asset, the risk of the individual assets selected as well as how much the prices of the assets move together. Details of the calculation will not be included here. Interested readers are referred to the literature. Readers should bear in mind that the method under the Markowitz model is popular, but not the only way to calculate portfolio risk. In Section 4.4, we described a different way to measure the risk of a portfolio used in the finance industry.

Figure 5.1 plots 10,000 random portfolios for three stocks in the London Stock Exchange from 22 January 2016 to 12 January 2018 (500 days); the three arbitrarily picked stocks are TSCO, BA and RBS.

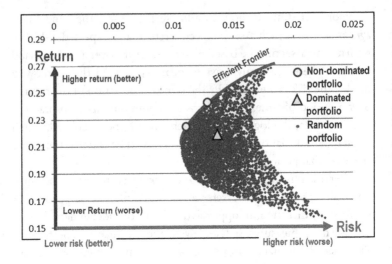

Figure 5.1 The Plotting of 10,000 Randomly Picked Portfolios for TSCO, BA and RBS, Three Stocks from the London Stock Exchange from 22 January 2016 to 12 January 2018 (500 Days) according to the Markowitz Model.

Each portfolio was created by allocating random weights to each of the three stocks. The risks (shown on the x-axis) and returns (y-axis) were calculated according to the Markowitz model.

The fund manager wants to maximize return and minimize risk. In computer science, this is a *two-objective optimization problem*. In Figure 5.1, both portfolios A and B (labelled in circles) are better than portfolio C (labelled in a triangle) because they provide a higher return and a lower risk than C. Portfolio C is said to be "dominated" by portfolios A and B. Instead of holding portfolio C, the portfolio manager will get a higher return and lower risk by holding either A or B. On the other hand, neither A nor B dominates the other, because A has a lower risk than B while B has a higher return than A. Among the randomly generated portfolios, the ones that lie on the top left quarter, labelled "efficient frontier" in Figure 5.1 dominate the portfolios below the frontier.

In practice, fund managers would reduce this problem to a single-objective optimization problem. They might determine the minimum return that it demands and pick the portfolio on the efficient frontier (hence with the lowest risk) that matches that return. Alternatively, they might determine the maximum risk that they are willing to bear and pick the portfolio on the efficient frontier (hence with the highest return) that matches the specified risk.

5.3 CONSTRAINED OPTIMIZATION

The Markowitz model assumes that one can allocate any weight to any asset in the portfolio. This implies that the fund manager can buy a fraction of a share. In reality, shares are bought in "lots", e.g. 100s. Sophisticated investors will be able to get around such constraints, but that complicates the operation and may not be cost-free.

The simplifying assumption impacts the computation: without constraints, the efficient frontier is smooth. That means once the fund manager finds a portfolio on the frontier, it can crawl along the frontier to neighbouring portfolios – crawling can be done by slightly adjusting the weights on some of the assets and assessing the adjustments' impact on risk and return. That way, it can find the portfolio that meets its minimum return or maximum risk requirements. What happens when shares must be bought in whole numbers or in lots? The efficient frontier will become ragged because a small adjustment in weight may not result in a whole number of shares in some stocks. That makes crawling much more difficult because not every point on the efficient frontier contains a valid portfolio. The problem becomes computationally much harder. The task of finding the optimal portfolio (to satisfy either the minimum return or maximum risk requirement) becomes intractable as it is haunted by combinatorial explosion (see Section 2.4).

Buying shares in whole numbers or in lots is just one of the many possible deviations from the Markowitz model. Other constraints may apply. For example, a fund manager may limit the percentage of funds to be invested in the same sector ("don't put all the eggs in one basket" principle). Another fund manager may limit the number of assets held in its portfolio so that it can watch these assets more carefully or limit its transaction costs when it needs to adjust its portfolio.

From a computational point of view, once these constraints are included, the nature of the problem changes further. The techniques best suited for finding portfolios vary depending on the nature of the constraints included. For example, if a large number of constraints are involved, then constraint satisfaction could be employed: The principle of constraint satisfaction is to use constraint propagation to eliminate areas where computation can be saved. In other words, while the constraints are the cause of the complication, they can be deployed to guide the search towards solutions.

In some problems, the fund manager may have some idea of whether certain assets are likely to form good portfolios. Such knowledge could be turned into heuristics in algorithms. Heuristics do not have to be correct all the time. But if they are correct more often than they are wrong, then they could help an algorithm to find better solutions more efficiently. Heuristic search is a very important area of research in AI.

To summarize, adding constraints to the Markowitz model could make the task of finding optimal solutions much harder or even intractable. In this case, knowledge of computation techniques could make a big difference. They could help a fund manager to find better portfolios faster than their competitors. Knowing what to buy or sell ahead of one's competitors is crucial to the success of a fund manager, as explained in Section 1.2.

5.4 TWO-OBJECTIVE OPTIMIZATION

The fund manager has two objectives: to maximize return and to minimize risk. However, in practice (as explained in Section 5.2), fund managers often solve the problem by focusing on one objective. This could be done in many ways, for example:

(1) Determine the maximum risk that the fund manager is willing to accept and find a portfolio with that risk, maximizing return.

(2) Determine the minimum return that the fund manager is willing to accept and find a portfolio with that return, minimizing risk.

(3) Combine the risk and return into one objective; for example, find a portfolio that maximizes the "Sharpe Ratio", which is the return minus the risk-free return divided by the risk.

Many investors would find the question: "how much risk are you willing to take?" difficult to answer. To start, it is difficult for ordinary investors to know how to quantify risks. Besides, the answer to this question depends on what return one is talking about. If the investors want to take as little risk as possible, but the portfolio manager can only find them a portfolio that returns 0.5%, then the investors may be willing to compromise and take on a bit more risk. Similarly, if the investors want to gain a return of 12%, but the best portfolio that the fund manager can find has a 50% chance of losing 90% of the capital, then the investors may be willing to reduce their return expectation accordingly.

Investors in general would find it difficult to determine their maximum risk or minimum return before the portfolio manager finds them a portfolio. In order to specify their preferences, investors need to know the trade-off between risk and return. In other words, to make their decisions, they would benefit from seeing the efficient frontier shown in Figure 5.1.

Given the above analysis, one should be motivated to treat portfolio optimization as a two-objective problem, as opposed to a single-objective problem. Multi-objective optimization is a well-established discipline in computer science. Therefore, it is a bit surprising that multi-objective optimization methods have not found their way into portfolio optimization in the industry.[2] This is probably because portfolio managers do not realize how mature the multi-objective research is and computer scientists with relevant expertise do not realize the opportunity of applying their techniques.

To treat portfolio optimization as a two-objective optimization problem, one does not attempt to find a single solution. Instead, one would attempt to find a set of non-dominating solutions on the efficient frontier. How many portfolios should the portfolio manager present to the investors? That is a question for the investors – they could ask for as many portfolios as they are willing to examine. Instead of determining how much risk to take and how much return to demand in advance, most investors would find it much easier to be given concrete portfolios to choose from. Comparing portfolios is much easier than making abstract decisions. For example, the investors may be given three non-dominating portfolios on the efficient frontier to choose from. Based on the investors' choice, the problem solver may generate two more portfolios on either side of the efficient frontier. This way, the investor's choice can be refined incrementally. Such refinement can be done by mature population-based multi-objective optimization methods.

5.5 THE REALITY IS MUCH MORE COMPLEX

So far, we have explained that the Markowitz model is a simple model for portfolio optimization. With simplifying assumptions, solutions are relatively easy to find. As constraints are added, the model better describes the portfolio manager's true considerations. But finding solutions becomes computationally more demanding. As we treat the portfolio optimization problem as a two-objective optimization problem (instead of reducing it to a single-objective optimization problem), specialized knowledge in computation is required. In this section, we are going to explain that reality is far more complex than what we have described so far.

To start, short selling is not allowed in the Markowitz model. In finance, short selling is an important strategy used by hedge funds. When short selling is considered, risk assessment is more complicated.

Little has been discussed in the literature that in reality, when portfolio managers maximize return, they should maximize return minus cost, where cost should include computation cost and the cost of acquiring expertise. For example, if simulations are used to calculate the portfolio risks (as described in Section 4.4), then fast computers will help to speed up the simulations – we shall come back to the importance of speed later. If the model is complex, expertise is required to compute good solutions and compute them fast. Computational expertise could be expensive to employ.

How much should the portfolio manager pay for computational expertise? In general, the better an expert, the higher it costs, but better solutions could be found fast. However, there is a problem: before employing an optimization expert, the portfolio manager would not know how much they can improve upon the current team. The potential improvement is not even easy to estimate. With so much unknown, it is difficult to determine how much to spend on computation expertise.

Another factor makes the reality even more complex: prices in the market will change and sometimes change fast. That means the optimal portfolio is time dependent. If an algorithm takes too long to compute the optimal portfolio, that portfolio would carry a different return and risk (because the prices of the assets have changed). The algorithm must watch the market while it conducts the computation. It must decide when to return a portfolio that the manager has a chance of acquiring. This is not the same as portfolio management – the problem of adjusting an existing portfolio following the change in asset prices – which is a separate problem.

The real portfolio optimization problem, even without portfolio management, is far more complex than what we have described in previous sections. None of these complications is described in textbooks. The problem is so complex that no computational methods have been developed for it; not even close. A portfolio manager who manages to model better (with as many of the considerations described earlier) and find solutions faster for a more realistic model will have an edge over its competitors. The competition is in research.

5.6 ECONOMICS VS COMPUTER SCIENCE

It is worth elaborating on the point that economists and computer scientists have very different views on portfolio optimization.

A typical computer scientist would say: give me the specification of the problem and I shall find you a solution. This is because, by training, computer scientists typically start with a specification of a problem. Given the specification, a competent computer scientist will pick the relevant techniques for solving the problem.

To the computer scientist, an economist would reply: specifying the problem is the whole of my research! Of course, they are right. The Markowitz model is just a rough approximation of the real problem that needs to be solved. Adding constraints to the specification, as explained above, is just one step closer to the real problem. Solving these approximated problems is not as important as finding a model that captures more of the fund manager's considerations.

Some economists may believe that once they can specify a problem, finding solutions is relatively unimportant. Most economists do realize that the perfect rationality assumption is unrealistic. Some would understand that finding the optimal solution is nontrivial, but they may believe that "if I cannot find the optimal solution, everyone else would have the same problem". Unfortunately, this is not true. Computer scientists would know that for intractable problems, one must settle for suboptimal solutions. Some methods will be able to find better solutions than others and find them faster.

An economist might say: "we are not dealing with a perfect model anyway. What does it matter if two solutions differ by 5% in the quality of their solutions?" The reality is that even if two methods can find solutions of similar quality, speed matters. By using the right algorithms and heuristics, one solver may spend a fraction of the time required by a naïve algorithm to find solutions of the same quality. For example, even the hardest Sudoku puzzle would just take a constraint satisfaction solver one to two seconds to solve on most home computers. A naïve brute force search would be haunted by the combinatorial explosion problem (explained in Section 2.4); it could take years to solve the same problem.

Speed matters especially if the fund manager needs to react to the market. So far, we have left out the problem of portfolio management (the problem of adjusting a portfolio to reflect price changes in the assets). To adjust a portfolio, the fund manager needs to know what the new optimal is. If it takes too long to calculate the new optimal, prices would have changed again. For that reason, computational speed matters.

The economists are correct in pointing out that in portfolio optimization, the research focus should be on the specification of the problem. But not many pay attention to the fact that given a specification, finding solutions is non-trivial. Different computational methods work well for different problems. Depending on the specification of the problem, different methods must be used. Not every computer scientist has the same knowledge about algorithms. Acquiring such knowledge is not free. Therefore, computational knowledge matters.

To summarize, in portfolio optimization, one needs to know what one wants to achieve – that is the task of specifying the problem. One also needs to know how to find good portfolios efficiently – that is the task of algorithms design and selection. To succeed in portfolio optimization, one needs synergy between finance and computation; one needs to know what methods work with what specifications. (Readers may refer back to Section 1.6, when we discussed synergy between finance and computation in trading.)

5.7 SUMMARY

Portfolio optimization is one of the core problems in finance. The fund manager has two objectives to achieve: to maximize return and to minimize risk. Diversification is the basic principle to reduce risk. Many practitioners start with the Markowitz model, for which optimal solutions are not hard to find, especially if one turns the problem into a single-objective optimization problem.

If one wants to better model the fund manager's needs, one must relax the simplifying assumptions and consider realistic constraints. This makes the problem harder to solve. The more factors the fund management considers, the more demanding it is for the fund management team's expertise in algorithms – they need to know what algorithms to use to handle what types of constraints.

There are a lot of opportunities for finance and computing experts to exploit if they could work closer together. The portfolio optimization problem is a two-objective optimization problem. Financial experts probably do not realize how mature multi-objective optimization research is; most computer scientists do not realize the potential of their techniques in this problem. Besides, if they both research deeper into the problem, they will realize that the current models ignore some of the most important aspects of portfolio optimization: computing expertise costs and computation takes time. How much computation expertise should a fund manager acquire? In a market where price change quickly, when should computation terminate to allow the fund manager to acquire a less-than-optimal portfolio? These are all open questions waiting to be studied.

> **The reality is messy. That is why we need to make models. But we must understand how our models compare with reality.**

NOTES

1. The return is the price change in percentage. Log return is often used by industry.
2. The author will not be surprised if some companies are using multi-objective optimization methods in portfolio optimization without publicizing it.

6

FINANCIAL DATA

BEYOND TIME SERIES

6.1 WHAT IS TIME EXACTLY?

Knowledge representation is an important part of AI. How one represents knowledge determines how one could reason about it. Nearly half of the research in early AI focused on knowledge representation, with the other half on searching methods, which included machine learning.

AI ≈ Knowledge Representation + Search

Recent research in AI may have underestimated the importance of knowledge representation. In this chapter, we shall look at the way that data is collected in finance and how it affects reasoning. We shall first look at the concept of time.

> "Time has no independent existence apart from the order of events by which we measure it". (Albert Einstein)[1]

DOI: 10.1201/9781003348474-6

Event-based logic has never been mainstream in time logics, but it was not ignored. Event–time logicians asked: if nothing ever happens, does time matter?

Time is extremely important in finance. Traditionally, price movements have been recorded by Time Series. Financial regulators, fund managers and traders pay attention to significant price movement events, especially when they happen within a short period of time. For example, the flash crash on 6 May 2010 was a talking point, because the Dow Jones Industrial Average dropped by about 9% within minutes, before it partially recovered within about half an hour's time.

The 2010 flash crash event will not be recorded on a Time Series that records daily closing prices. This event will not be fully recorded on an hourly recorded Time Series either. It will only feature in a minutely recorded Time Series. One must then ask: with Time Series, what is the right frequency to record transactions in a market?

Before answering that question, one must understand that observers such as financial regulators, fund managers, traders and investors are interested in events. They look at the 2010 flash crash as a crash event followed by a recovery event within a short period of time. In Time Series, prices are recorded at fixed intervals. Events in Time Series are only *secondary objects*, to be derived from prices and times recorded. Arguably, there is a mismatch between the observers' interest and the way that prices are recorded in a Time Series. By observing the market with daily closing prices, for example, the flash crash would not have been observed, because the market would have recovered by the time the next data point is recorded. If secondly transactions are recorded, then the observer will have to deal with a lot of data which is uninteresting.

If events are what observers pay attention to, there is no reason why one should not record events as *primary objects*. That means, instead of recording transactions at fixed intervals, one could directly record the events that happened in the market. This motivates the concept of "Directional Changes", which will be described in the next section.

6.2 EVENT-BASED TIME REPRESENTATION

Directional Change (DC) is an event-based representation. Instead of recording transactions at fixed intervals, as it is done in Time Series, DC focuses on 'Directional Change Events' in the market. Before the observer starts, it must determine how big a price change is significant. Different observers view the market differently: A long-term investor may consider 10% as a significant change, but a day trader may consider a 0.2% change significant. We shall refer to this percentage as a "DC threshold". Depending on the DC thresholds used, different observers may see different pictures of the same market that is useful for their individual purposes, as is the case in Time Series where different observers choose to record transactions at different time intervals.

In DC, a transaction is only recorded when the price moves in the opposite direction of the current trend by the DC threshold specified. Suppose an observer uses 5% as its threshold. Suppose further that the current trend is going up. If one observes a transaction which price is 5% below the highest price of the current trend, then one records a Directional Change Event in the market. From then on, the current trend is recorded as going down. The observer will know that the downtrend has ended when it observes a price which is 5% above the lowest price of this downtrend. The market is therefore recorded as a sequence of alternating uptrends and downtrends.

Traders are familiar with the terms "bull" and "bear". A bull market is a market in which prices tend to go up. A bear market is one in which prices tend to go down. DC can be seen as a formal definition of bull and bear.

Figure 6.1 shows an artificial data set. The horizontal axis shows the time and the vertical axis shows the price. Each circle represents a transaction. Transactions take place at irregular times. Each square represents a recorded transaction under Time Series. As transactions do not take place at fixed intervals (which are indicated by the vertical lines), most of the squares do not overlap with the circles. Transactions recorded in Time Series are only based on the

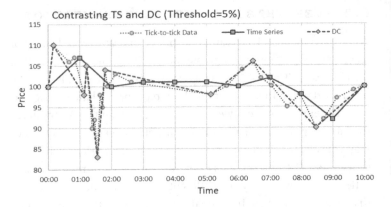

Figure 6.1 Contrasting Time Series and Directional Change Sampling.

transactions before the sampling time. As no transaction took place between 03:00 and 05:00, missing data points must be artificially constructed for Time Series. In this example, the preceding transaction price 101 (at time 02:38) is copied for time 03:00, 04:00 and 05:00 in the Time Series.

Each diamond in Figure 6.1 represents a data point recorded in DC. At those points, the market changes direction. For example, from 00:10 to 01:08, the market was in a downtrend because the price has dropped more than 5% from 110 (at 00:10) to 98 (at 01:08). An uptrend was found from 01:08 to 01:13 because the price has risen by more than 5% from 98 (at 01:08) to 105 (at 01:13).

One important point to note is that each recorded transaction in DC is an actual transaction (unlike Time Series, which are approximations and sometimes artificially created). Like the raw data, DC records transactions at irregular times. For that reason, most analyses in Time Series do not apply to DC series. A new representation demands new reasoning methods. While this poses new challenges, it also offers new opportunities, as we shall explain later.

The substantial price drop followed by a price reversion of similar magnitude between 01:00 and 02:00 in Figure 6.1 represents a flash crash. By recording transactions at fixed intervals, Time Series misses those activities – it will only record the price of the final transaction within each interval. On the other hand, this flash crash will be recorded under DC, as it records all significant price changes according to the DC threshold. Between 01:00 and 02:00, four extreme points (points at which direction changes) were recorded. In other words, the direction has changed four times within that period.

It is important to note that Directional Change Events are recognized in hindsight. The transaction at 01:13 (at price 105, see the fourth diamond from the left in Figure 6.1) was only recognized as a DC point when the transaction at 01:23 (at price 90, see the grey circle in Figure 6.1) was observed, as the price 90 is more than 5% below 105.

6.3 MEASURING MARKET VOLATILITY UNDER DC

How one records prices in the market determines what one can reason about. Starting with the same set of transactions, Time Series and DC select different transactions to record. How would they reason about data differently? We shall focus on risk measures under DC.

In Chapter 5, we explained one way to measure risk under Time Series: the log return of each period is calculated. The risk for a period can be measured by the standard deviation of the corresponding series of log returns. How would risk be measured under DC?

One way to measure risk in a period under DC is to count the number of direction changes in that period. In Figure 6.1, we can see four directional changes between 01:00 and 02:00, but no directional changes between 02:00 and 05:00. This shows that the market is very volatile from 01:00 to 02:00 but not volatile at all from 02:00 to 05:00. In general, the frequency of directional changes in a period is a simple way to measure volatility for that period. The higher the frequency of directional changes, the more volatile that period is, which means the more risky it is to trade in that period.

Another way to measure the volatility of the market under DC is to measure the magnitude of price changes from the start to the end of each trend. For example, in Figure 6.1, within the period (01:08–01:48), three trends were completed. Directional changes were recorded at the prices of 98, 105, 83 and 104. In the first of these three trends, prices have changed by (105−98=) 7. In the second and third trends, prices have changed by (83−105=) −22 and (104−83=) 21. If we take the absolute values of these changes, the total price changes are (7+22+21=) 50. The average magnitude of the price changes in three trends in this period is, therefore (50/3=) 16.7. Without going into details, the magnitude of price changes in the next three trends from 01:48 to 08:28 averaged 10. The values 16.7 and 10 are quantitative measures of the volatility of the two periods (01:08–01:48) and (01:48–08:28) under DC. The former is more volatile than the latter under this measure.

To summarize, we have introduced two quantitative measures of volatility under DC:

(1) The frequency of direction changes.
(2) The average magnitude of price changes per trend.

These two measures are orthogonal: directions could change frequently in a market, but the magnitude of price change in each trend could be small, or it could be big. The same is true for a market in which directions may change infrequently. So both measures are useful for describing the volatility of a market.

Above are just two examples of measuring volatility under DC. More have been defined, which will not be included here. The important point is that these two measures add to the standard deviation of log returns under Time Series to give the observer multiple perspectives in risk analysis.

6.4 TWO EYES ARE BETTER THAN ONE

One may ask: why should one bother looking at DCs? We have been using Time Series happily. We know how to handle Time Series.

We know how to extract information from them. Software packages are available for dealing with them. There are plenty of published results to compare our results with. We know how to interpret our results. Why change?

The short answer is: DC is not a replacement for Time Series, but an additional tool. In a financial market, being able to extract more information from the data gives one an edge over one's competitors. Under DC, one may be able to observe things that cannot be observed under Time Series. We have explained above that by recording the extreme points, activities in the market will not be missed under DC. For example, the price changes between 01:00 and 02:00 in Figure 6.1 will not be recorded in a Time Series that records one transaction per minute, but they will be recorded in a DC summary. Being able to record these price changes gives the observer a chance to reason about them.

We also explained in the previous section that volatility can be measured under DC using (1) the frequency of DCs and (2) the average magnitude of price changes per trend. These observations are orthogonal to volatility measures using the standard deviation of log returns under Time Series. The volatility measures under DC and those under Time Series may or may not agree with each other. By using both DC and Time Series, one may be able to see things that one could not see with just one of them.

Time Series and DC start with the same set of raw transactions from the market. Their difference is in the way that they choose transactions to record. Time Series records the transactions at fixed intervals (by taking the final transaction before the sampling time). DC records the transactions at which significant changes take place. Selecting raw transactions to record is necessary for analysis. But either way to select transactions is perfect. Using Time Series and DC together reduces the chance of missing blind spots. What Time Series and DC see are just two different views of the same market over a period. The two views may agree with each other. But occasionally they do not, which may tell us something that other observers fail to see.

Seeing with two eyes is often better than seeing with one. One series may reveal information that is not captured by the other. For example, DC captures a flash crash between 01:00 and 02:00 in the market, which is not captured by the corresponding Time Series. There are statistical measures captured by Time Series which are not captured by DC.

In Chapter 3, we emphasized the importance of data in machine learning. Above, we have introduced the frequency of directional changes within a period as a measure of volatility in the market. This measure is independent of the volatility measures under Time Series. More new volatility indicators have been defined under DC. By sampling transactions differently, DC and Time Series introduce different sets of variables. Machine learning relies on data. New variables measured under DC have created new opportunities in machine learning. Practitioners equipped with more variables will have a better chance to beat their competitors in forecasting and risk analysis.

Going back to the question at the beginning of this section: the question is not about moving from Time Series to DC or whether DC is better than Time Series. It would be foolish not to use all the information that one can get one's hands on in a competitive market. One should always use both DC and Time Series, plus any new representations of time that may be developed in the future.

6.5 STRIKING DISCOVERIES UNDER DC

The most important stylized facts under DC were observed by Olsen and his team in the foreign exchange market. They observed the foreign exchange markets across all major currency pairs over a long period. They discovered very interesting statistics. across a wide range of DC thresholds. The following are two of the most striking discoveries:

- If 5% is the DC threshold, then on average a trend ends when it reaches 10%.
- If a trend takes 1 minute to reach the DC threshold, then on average it takes another 2 minutes to reach the end of the trend.

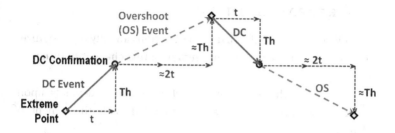

Figure 6.2 Striking Observations in the Forex Market under DC: Across All Currency Pairs, Across All DC Thresholds (Th), (1) On Average, the Trend Ends When it Reaches Twice the Threshold; (2) On Average, if a Trend Takes a Certain Amount of Time (t) to Reach the Threshold, It Takes Twice as Much Time (2t) to Finish.

These observations are summarized in Figure 6.2. For convenience, we call the price change from an extreme point (indicated in diamonds in the figure) to one threshold a DC event. The price change from the DC confirmation point (at which the price has changed by one threshold in the opposite direction of the previous trend) to the next extreme point is referred to as the Overshoot (OS) event.

Mathematically minded readers should note that the average is significantly biased by extreme values. That means the majority of trends ended far sooner than reaching twice the threshold value but extreme trends ended much later (say, six or seven times the threshold). Details of this observation are still under ongoing research.

These were observations; no explanation is available. What are the implications of these observations? The implications are still under research. Could they help traders to develop trading strategies? It is up to traders to find out.

As these results have been published, traders will find ways to exploit them. Traders who find ways to exploit such statistics ahead of their competitors will benefit. When enough trades have exploited them, these statistics are likely to disappear from the market. It is up to researchers to find new stylized facts. As we explained in Section 1.3, "it takes all the running you can do, to keep in the same place". The competition is in finding regularities ahead of one's competitors.

6.6 RESEARCH IN DC

What is the use of DC if a change of direction can only be recognized in hindsight, one may ask? The answer is: that changes nothing.

- For data in the *past*, which machine learning depends upon, extreme points in DC can be recorded with the benefit of hindsight.
- Whether one is using DC or Time Series, one can only reason with data up to the *present*. No matter how one collects data, one could only know whether the market has turned from bull to bear when the price has fallen deep enough. Similarly, only after seeing enough price rises could one conclude that a bear market has turned bull.

For the above reasons, the fact that a change of direction can only be recognized in hindsight affects neither one's analysis of the past nor the present.

One could forecast under DC, as one does under Time Series. Research has been conducted in forecasting whether an uptrend (a bull market) will reach a certain height, or a downtrend (a bear market) will reach a certain low. Like forecasting under Time Series, forecasting under DC uses machine learning. In financial forecasting, machine learning starts with data. Historical data are used to look for hints of price movements. As Time Series and DC use different indicators (we introduced two DC volatility indicators above), their results provide independent forecasts, which could be used to verify each other. If forecasts under Time Series and DC agree with each other, we should have more confidence in the forecasts. On the other hand, if the results contradict each other, we may have to be more cautious in using the forecasts. Two eyes are better than one, as we suggested above.

It is worth introducing the "nowcasting" problem in DC. As explained above, the end of the previous trend in DC is only confirmed when we see a price reversion by the DC threshold. Before this threshold is reached, the observer is unaware that the market is in a new trend. Can one detect that a new trend has already started? This is called a "nowcasting" problem because it is trying to detect what has already happened (as opposed to forecasting the future).

As is the case in forecasting, nowcasting involves using machine learning. With past data, it uses DC indicators to look for signs of the previous trend finishing. To give the readers an idea of how this can be done, here is where hints may come from: suppose we use a DC threshold of 5%. Suppose we are in a DC downtrend and the price has dropped from the previous extreme point by 30%, which is six times over the threshold. According to the stylized fact introduced in the previous section, on average, a trend ends after it reaches twice the threshold – which is 10% in this case. This suggests that the current trend may end soon if it has not already ended. On top of that, suppose the current transaction has reversed by 4%, then one may have good reasons to suspect that a new trend has already started from the lowest point of the current trend. Obviously, this guess may be wrong, but it is not a bad guess. The point here is to explain that hints could be found for nowcasting in DC.

With DC indicators, one could monitor the market for abnormality. One piece of such research is "regime change" detection. A regime change is said to have happened if the market enters a state in which the statistical properties of price changes differ from what was normally observed. This research starts with data. Machine learning is used to learn models based on indicators defined under DC. These models are used to monitor the market transaction by transaction, which reveals probabilities of whether a regime change has taken place. Being able to monitor regime change is important to traders. A trader may want to adopt a different trading strategy when the market regime has changed. Alternatively, it may choose to close its positions[2] when the regime has changed. Regulators may want to monitor the market tighter when the regime changes, in case the new regime leads to extreme turbulence.

6.7 CONCLUSION: NEW REPRESENTATION, NEW FRONTIER

Financial market dynamics are traditionally recorded in Time Series. What is the best frequency to record a transaction? Daily? Hourly? Minutely? An hour is a short time in a sluggish market, but during a flash crash, one second could be a long time. The best approach is to look at the market from a different angle: let events dictate when to record a transaction. This motivates the definition of Directional Change (DC), an event-based representation of time.

DC provides an alternative way to Time Series in transaction data collection. It provides one with more information about the market, such as new measures in volatility (Section 6.3). With the same raw transaction data collected differently, one sees the market from a different angle (Section 6.4). This allows one to see things that one could not have seen before; the stylized facts observed in the foreign exchange market are good examples (see Section 6.5).

Most researchers are comfortable with Time Series. Is there any incentive to look at DC as well? The answer is yes. When everyone looks at the same place, all the low-hanging fruits will be gone. DC provides a new perspective to researchers. With this new perspective comes new opportunities. More importantly, looking with two eyes is likely to be better than looking with one.

Machine learning (especially supervised learning, see Section 3.2) starts with data. How one collects data determines what one can reason about. Data collected for DC fuel machine learning for discoveries independent of Time Series. Stylized facts observed under DC (Section 6.6) fuel new research too.

As a new representation, DC research demands new reasoning methods. The new representation provides opportunities to those who know how to interpret and analyze DC series. DC research is in its infancy. Thousands of person-years research must have been put into Time Series. Research in DC is probably in tens of person-years. Many low-hanging fruits are waiting to be picked.

NOTES

1. L. Barnett, *The Universe and Dr Einstein*, Dover Publications, Inc., 1985 (p. 19).
2. Closing a position on an asset means selling any holding and paying back any borrowing of the asset.

7

OVER THE HORIZON

7.1 ALGORITHMIC TRADING DRONES

Most trades are conducted by computer programs today. Human trading will become a rarity in the future. There will always be human traders who have insight into how to beat the market, but most of the established trading strategies will be implemented in computer programs. Machine learning will invent trading strategies beyond human traders. Human trading will still take place in specialized markets and less active markets for which returns may not justify the investment of algorithmic trading.

Human traders may be smarter than computer programs at places, but they cannot take input and react as fast as computer programs. They need rest, which means they may miss opportunities. Computer programs do not need to eat, drink or rest. They can pay full attention to multiple markets 24 hours a day. They can react much faster than human traders.

Costs will play a part in the fading out of human traders too. Human traders are expensive to use – they need to be paid. In contrast, once implemented, computer programs belong to the companies that paid for their development. Computer programs can be replicated. That means if the logic is proven to be good, the same program can be tested in multiple assets and markets. Human traders, no matter how clever they are, cannot watch multiple markets

DOI: 10.1201/9781003348474-7

at the same time. Their best hope is to channel their expertise into computer programs and let the programs trade on their behalf.

Reliability favours algorithmic trading too. Computer programs do not have emotion. They will not panic when the market goes against their trades. They will not try to recoup their losses by taking unnecessary risks; human traders may do so to save their jobs. Computer programs can be audited. When serious mistakes are made, the culprit codes that led to big losses can be removed or rewritten so that the same mistakes can be avoided.

Computer programs accumulate the expertise that contributed to their development: When human traders leave a company, they leave with their expertise. But if the company managed to channel the traders' expertise into a computer program, the program will accumulate expertise from multiple traders. Hence, the program will get better and better over time.

Algorithmic trading is already a major player in major financial markets. So much so that anti-machine trading algorithms have been developed. "Spoofing" is a good example. Spoofing programs place bids which are way below the current price or offers which are way above the current price in the market, only to be withdrawn within milliseconds. These orders are unlikely to be executed because they are withdrawn before the price moves. The purpose of spoofing is to mislead other computer programs. Only computer programs will be able to notice orders that are placed and withdrawn within milliseconds. The spoofing programs aim to trick the other trading programs into believing that there is huge demand or supply in the asset. Their aim is to move the price in the direction in their favour.

Machine learning has been used to detect spoofing orders. If spoofing can be recognized, it can be ignored. If we think of the algorithmic trading programs as drones, then the spoofing programs are anti-drone drones. The spoofing recognition programs are therefore anti-anti-drone drones. The arms race in algorithmic trading is heating up.

Algorithmic trading has been blamed for causing crashes in the markets. Indeed, programming bugs and spoofing activities may cause undesirable movements in the market. However, this can be prevented in many ways. For example, the regulator could require all programs that trade above a certain volume to go through tests. If drivers are required to acquire a license before they can drive vehicles, why should programs not be required to acquire a license before they can trade, given that misbehaved programs could potentially wipe out millions of pounds/dollars from a market? Some of the tests could subject the trading program to past turmoil market situations to see whether it causes more turmoil or the reverse.

In fact, with increased transaction frequency, algorithmic trading will provide liquidity to the market. Therefore, when properly regulated, algorithmic trading should lower volatility in the market, not the opposite.

7.2 HIGH-FREQUENCY FINANCE

High-frequency finance refers to financial activities that use high-frequency data and trade at high frequency. High or low frequency is a relative concept. In an active market, such as the euro–dollar exchange market, which is a 24-hour market, recording one transaction per day is pretty low in frequency. Recording one transaction per hour is higher in frequency. The limit is to record and use every transaction in the market.

The following example should explain why high-frequency finance is important: suppose a trader inspects the price of an asset once every day. Suppose the price of this asset rises from 100 on Day 1 to 110 on Day 2. If the trader successfully predicts this rise on Day 1, they could buy at 100 and sell at 110, gaining a 10% profit. Now suppose this trader inspects the market three times a day (whether the inspections take place at equal intervals does not affect this analysis). Suppose the price changed from 100 to 107 and 103 before it reaches 110 on Day 2. If this trader manages to predict these changes, then they could have bought at 100, sold at 107 and bought back at 103 before selling at 110. It would have gained 7% in the first trade and 6.8% in the second trade.[1] Together, ignoring compound interest, it would have gained 13.8%.

This example shows that by inspecting the market more frequently, the trader could potentially gain a higher profit (13.8% vs 10%). Of course, the trader may not be able to predict the prices accurately. But if this trader were to invest their time in studying the market with the goal to forecast price changes, inspecting the market at a higher frequency is a simple operation to increase their potential in gaining a higher return.

When the market was inspected once per day, the price change in the above example was (110−100=) 10. But if the market was inspected three times on that particular day, the price changes were +7, −4 and +7. Taking the sign away, the total price change was (7+4+7=) 18. The more frequently one inspects the market, the bigger the total price change one would find.

The analogy is in the measuring of the length of a coastline (following Mandelbrot): a coastline may look smooth from a satellite. As one descends to, say, 2,000 metres, one can see more details (such as the mouth of a river), hence a longer coastline. When one descends to sea level, the coastline will measure even longer.

If high-frequency data could potentially help a trader make more profit, then why should anyone not use them? Here are some deterring reasons: firstly, not everyone has access to high-frequency data. Data feed costs. Data storage costs too. Secondly, not everyone knows how to make use of them. It has been argued that directional change (DC, see Chapter 6) is more suitable to process high-frequency data than Time Series, but research in DC is still in its infancy. Finally, human beings cannot react to market changes in microseconds. High-frequency trading can only be done by algorithmic trading (Section 7.1), which itself demands investments and expertise.

As more people gain access to high-frequency data and know how to analyze them, high frequency will become more popular. Researchers who use high-frequency data before others will be able to harvest the low-hanging fruits before their competitors.

7.3 BLOCKCHAIN

The best way to understand blockchain is to detach it from bitcoin, which is often associated with it. Bitcoin is a cryptocurrency which uses blockchain to support its transactions. Blockchain is the underlying technology which can be used for transactions other than bitcoin.

Blockchain is just a ledger in a bookkeeping system. It records who owns what, just like a bank recording how much money is under which account. The main difference is that a blockchain is a ledger that makes many copies. Anyone who is involved in the transactions may keep a copy of this ledger.

What is the significance of having multiple copies? That makes forgery difficult. One may be able to change one's own copy of the ledger. When the next transaction takes place, the system will detect that the two ledgers do not match.

Being hard to forge makes blockchain very useful for the trust business – a role played by banks, credit card companies, PayPal, Apple Pay, Google Pay and other payment systems today. Note that blockchain is not a competitor to these payment services. Instead, it is a ledger system that can be used by these services. By looking after the booking, blockchain helps new services to be established in the trust business.

Whether blockchain will be widely accepted by businesses and individuals depends on many factors, including regulations, cost and the public's perception of it. But if accepted, it has the potential to disrupt the trust business.

As blockchain is a system for recording who owns what, it can be used to record the ownership and transactions of normal currencies. Platforms have been started to trade currencies as well as cryptocurrencies using blockchain.

It does not have to stop there. One could, for example, use blockchain to record who owns what shares and how much. Once shares are recorded using blockchains, stock exchanges could clear transactions with blockchains. Blockchain could also be used by the Land

Office to record the ownership of properties. In fact, it can be used to record the ownership of anything, from goods to personal belongings.

Potentially, blockchain enables platforms to be established to trade anything. Speed in clearing and reduced risk make blockchain attractive to users: Assets whose ownership is recorded under the same blockchain can be exchanged instantaneously. With electronic contracts, credit risk is reduced because ownership will be exchanged through automation after the transactions. Operational risk is low too.

Blockchain is not a result of AI research. However, with improved efficiency and security, blockchain makes it easy to set up new markets. Modelling, simulation and machine learning are particularly useful for designing market rules (see mechanism design in Section 4.5). Besides, with electronic contracts, all the terms and procedures must be clearly specified. With formally stated specifications, inferences can be made. Automated deduction (a branch of "good old AI" which is still relevant though not fashionable) will become possible.

7.4 INFORMATION EXTRACTION FROM NEWS

We emphasized the importance of data and the importance of knowing the data in machine learning (Section 3.3). The key data in finance are transactions in markets. But that is not the only source of data. The prices exhibit the results of the traders' collective behaviour. The traders' decisions are influenced by their confidence, which is in turn influenced by news and opinions. Texts from news pieces and social media such as Twitter can be fed into computer programs as data. From these data, information can be extracted.

One branch of research that is growing and has a lot more scope to grow further is in extracting information from texts. Programs have been developed to take news feeds and social media feeds as their input. By processing such data, they output the moods of the market. From a piece of news, some programs may output a simple conclusion classifying whether it is positive or negative. Some programs may output a mood indicator on a scale.

The Federal Reserve issues reports on interest rates periodically. These reports have a significant impact on markets, especially the foreign exchange markets. Given the significance of these reports, specialized programs have been developed to read these reports to micro-study the wordings in order to extract information from them.

As news and texts in social media are written in natural language, these programs must be able to "understand" natural language. Natural language understanding is an important branch of AI. Research in sentiment analysis attempts to extract from news texts the mood of the market. Digested data can be fed into computer programs for machine learning (Chapter 3), risk analysis (Section 4.4) and portfolio optimization (Section 5.5) and algorithmic trading (Section 7.1).

Information extraction from texts is not straightforward. Recognizing keywords alone is not enough. The programs must take into consideration many human factors. For example, bad news tends to get reported and twitted more. News will not report business as usual. Besides, when a company is in financial trouble, it often makes announcements which emphasize its financial stability; sometimes this is done through its influence over newspapers or opinion leaders. Researchers must take these into consideration.

Investors, traders, fund managers and regulators could all benefit from information extracted from more sources of data. No one can pay attention to all the news and social media. But computer programs can read from newspapers, tweets and other social media day and night. The potential of sentiment analysis has not been fully realized yet.

7.5 FINANCE AS A HARD SCIENCE

Every aspect of a market can be run by computers. This includes order clearing in the stock exchange, market making in foreign exchanges, algorithmic trading and electronic contracts.

Imagine a market in which all the programs start from formal programming specifications,[2] and all the programs are automatically generated from specifications.[3] If automated programming does its job, then all these programs will do exactly as intended. When this is the case, we can study the behaviour of these programs rigorously, as in mathematics and logic. Markets can then be studied rigorously, just like how we use physics to study the natural world. Arguably markets should be easier to study than the natural world because all computer systems are human-made, so we should know exactly how they work. Therefore, when all processes are rigorously specified and implemented correctly in an automated market, properties of the market could be studied as hard science. In this hypothetical world, experiments can be repeated. Control experiments can be conducted.

This is not to suggest that we shall fully understand what will happen in this hypothetical world. Firstly, we may know the programs, but we do not know the data. We do not know how people decide to buy and sell and at what prices. We do not know how much money individual investors will have. We do not know what margins traders will use to trade with. Secondly, it is a complex system. Complex systems are hard to study, even if we know all the causal relations within them.

There is a big gap between the current situation and the hypothetical world sketched above. Most computer program specifications are written in natural language, not formal specification languages. Natural language can sometimes be ambiguous. Program implementations are rarely bug-free. Programming bugs are generously tolerated; people rarely will go beyond moaning when they encounter operating system failures ("blue screens"), for example. Traders generally accept "glitches" caused by programming bugs. The order-clearing algorithms are not necessarily transparent. Dark pools are accepted by market participants. Market-making algorithms in setting bid and offer prices are not normally disclosed. Regulators have access to a lot of data, which allows them to conduct stress tests. But they are still far away from being able to study markets like physics, where control experiments and repeatable experiments are expected.

Will that hypothetical world emerge? Maybe, but more likely not. That depends on the collective wills of all the parties involved. The point is: the more imperfect the markets, the more mispricing in assets. Mispricing is buried in complex systems. This is the best time for mispricing seekers! Knowledge of AI helps.

> *As long as the market is not a hard science, misbehaviour in markets is common. Knowledge of AI helps in exploiting such opportunities.*

NOTES

1. For simplicity, we ignore the possibility of short-selling, which could have gained the trader more profit.
2. Program specification is a field in computer science. The idea is to use some formal languages to unambiguously describe propositions or functions.
3. Automated program generation from specifications is a field of computer science (sometimes classified under AI).

BIBLIOGRAPHICAL REMARKS

AI for finance is an evolving subject. Literature on AI for Finance is relatively scarce. The latest techniques are kept in companies; often, they do not tell others (especially their competitors) what they are researching on. No textbook is available to beginners.[1] Such books have not been written yet because the scope of this subject is not yet defined. The fact that different readers have different needs makes such books harder to write. Some need more background in computing and others more in finance. Frontier research is published in the form of technical papers, which tend to be difficult to read.

AI texts are abundant. *Artificial Intelligence: A Modern Approach* by Russel and Norvig (2021) is comprehensive and covers the most important areas in AI. Readers who want to know the scope of early AI should consult *The Handbook of AI* (1981–1982). *Pattern Recognition and Machine Learning* by Bishop (2007) is an excellent text on neural networks for machine learning. Consult Tsang (1993) for a formal introduction or Rossi et al. (2006) for the full scope of constraint satisfaction problem-solving.

General finance and economic texts are abundant. Not a few references could cover all major areas. We mentioned momentum trading in algorithmic trading (Chapter 1); readers interested in technical trading may consult Krausz (2006). Arbitrage opportunities

(Section 1.2) in the London Stock Exchange were spotted by Tsang et al. (2005). With an extensive examination, Faleiro and Tsang (2016) show that momentum trading strategies are no longer reliable in today's markets.

AlphaGo (Section 2.1) and AlphaGo Zero (Section 2.2) ignite the public's interest in AI. They were reported by Silver et al. (2016, 2017).

Tsang and Li (2002) explain how genetic programming could be applied to forecasting (Sections 3.1 and 3.2). The idea is extended by Kampouridis et al. (2012, 2013). Tsang et al. (2005) explain how the idea is applied to arbitrage forecasting. Genetic programming is extended by Garcia Almanza and Tsang (2011). The Basic Alternating-Offers Model (Section 3.4) was discussed by Rubinstein (1982); *Bargaining Theory with Applications* by Muthoo (1999) is an excellent book on bargaining theory. Jin and Tsang (2011) explain how a constraint-directed genetic programming approach could be used to find subgame equilibriums. GPBIL (Kern 2005) is arguably the simplest machine learning method which has been applied to finance (e.g. see Alexandrova-Kabadjova et al. 2011).

Farmer and Foley (2009) argue for the modelling in economics. Alexandrova-Kabadjova et al. (2012, 2015) collect important research on simulation applied to payments in central banking policies on interbank payments (Section 4.1). Garcia Almanza et al. (2012) explain how genetic programming can be used to predict bank failure. Marquez Diez Canedo and Martinez-Jaramillo (2009) use modelling to study systemic risk in the banking system.

The Markowitz model for portfolio optimization (Chapter 5) is described in many publications, including Wikipedia. Zhang et al. (2010) addressed the portfolio optimization problem with constraints (Section 5.3). Saini and Saha (2021) survey multi-objective optimization (Section 5.4). A survey is incomplete without attention paid to MOEA/D, a state-of-the-art method by Zhang and Li (2007) which is well summarized by Li (2021).

The concept of directional change (DC, Chapter 6) was invented by Richard Olsen (see Dacorogna et al. 2001). A similar idea was introduced as "zig-zag" in technical analysis (Sklarew 1980), which lacks follow-up research. The striking discoveries (Section 6.5) were reported by Glattfelder et al. (2011) and Bisig et al. (2012). *Detecting Regime Change in Computational Finance, Data Science, Machine Learning and Algorithmic Trading* by Chen and Tsang (2021) is the most comprehensive book on this topic; it describes the regime change detection research mentioned in Section 6.6. Tsang (2021) argued that DC is suited for tick-to-tick data. Readers who are serious about event-based time may consult Van Benthem (1983, Chapter I.5).

Dempster and Leeman (2006) describe an automated FX trading system (Section 7.1). Golub et al. (2017) describe an algorithmic trading algorithm based on Directional Change (Chapter 6). Cao et al. (2015, 1016) describe how price-manipulating trades can be detected in algorithmic trading drones' warfare (Section 7.1). *An Introduction to High-Frequency Finance* by Dacorogna et al. (2001) is the best introduction to high-frequency finance (Section 7.2). The coastline analogy was invented by Mandelbrot (1982). Tsang (2021) argues that directional change is more suitable for handling high-frequency data than Time Series. See *The Handbook of Artificial Intelligence* by Barr et al. (Volume 3, 1986) for automatic deduction in AI (mentioned in Section 7.3). Tsang et al. (2013) show how liquidity risk can be inferred (not forecasted) when the market clearing mechanism is formally specified and order queues information is available (Section 7.5). *The Fractal Geometry of Nature* by Mandelbrot and Hudson (2004) is a good read on the misbehaviour of markets. *Inefficient Markets: An Introduction to Behavioral Finance* by Shleifer (2000) is an excellent text on market inefficiency and behavioural finance.

NOTE

1. The author does not consider this book a textbook. It is an easy read on the subject.

BIBLIOGRAPHY

Alexandrova-Kabadjova, B., Tsang, E.P.K. & Krause, A., Market structure and information in payment card markets, International Journal of Automation and Control (IJAC), 8(3), 2011, 364–370.

Alexandrova-Kabadjova, B., Martinez-Jaramillo, S., Garcia-Almanza, A.L. & Tsang, E.P.K. (ed.), Simulation in Computational Finance and Economics: Tools and Emerging Applications, IGI Global, 2012.

Alexandrova-Kabadjova, B., Heuver, R. & Martínez-Jaramillo, S. (ed.), Analyzing the Economics of Financial Market Infrastructures, Business Science Reference, 2015.

Barr, A., Feigenbaum, E. & Cohen, P. (ed.), The Handbook of Artificial Intelligence, Volumes 1 to 3, 1981–1982, reprinted by Addison Wesley, 1986.

Bishop, C., Pattern Recognition and Machine Learning, Springer-Verlag, 2007.

Bisig, T., Dupuis, A., Impagliazzo, V. & Olsen, R.B., The scale of market quakes, Quantitative Finance, 12(4), 2012, 501–508.

Cao, Y., Li, Y., Coleman, S., Belatreche, A. & McGinnity, T.M., Adaptive hidden markov model with anomaly states for price manipulation detection, IEEE Transactions on Neural Networks and Learning Systems, 26(2), 2015, 318–330.

Cao, Y., Li, Y., Coleman, S., Belatreche, A. & McGinnity, T.M., Detecting wash trade in financial market using digraphs and dynamic programming, IEEE Transactions on Neural Networks and Learning Systems, 27(11), 2016, 2351–2363.

Chen, J. & Tsang, E.P.K., Detecting Regime Change in Computational Finance, Data Science, Machine Learning and Algorithmic Trading, CRC Press, September 2021.

Dacorogna, M.M., Gencay, R., Muller, U., Olsen, R.B. & Pictet, O.V., *An Introduction to High-Frequency Finance*, Academic Press, 2001.

Dempster, M.A.H. & Leemans, V., An automated FX trading system using adaptive reinforcement learning, *Expert Systems with Applications*, 30(3), 2006, 543–552.

Faleiro, J.M. Jr & Tsang, E.P.K., Black magic investigation made simple: Monte Carlo simulations and historical back testing of momentum cross-over strategies using FRACTI patterns, working paper WP078-16, Centre for Computational Finance and Economic Agents (CCFEA), University of Essex, November 2016.

Farmer, J.D. & Foley, D., The economy needs agent-based modelling, *Nature*, 460(7256), 6 August 2009.

Garcia Almanza, A.L. & Tsang, E.P.K., *Evolutionary Applications for Financial Prediction: Classification Methods to Gather Patterns Using Genetic Programming*, VDM Verlag, 2011.

Garcia Almanza, A.L., Martinez Jaramillo, S., Alexandrova-Kabadjova, B. & Tsang, E.P.K., Using genetic programming systems as early warning to prevent bank failure, Chapter 14, In A.Y. Yap (ed.), *Information Systems for Global Financial Markets*, IGI Global, 2012, 369–382.

Glattfelder, J.B., Dupuis, A. & Olsen, R., Patterns in high-frequency FX data: Discovery of 12 empirical scaling laws, *Quantitative Finance*, 11(4), 2011, 599–614.

Golub, A., Glattfelder, J. & Olsen, R.B., The alpha engine: Designing an automated trading algorithm (April 5, 2017). High Performance Computing in Finance, Chapman & Hall/CRC Series in Mathematical Finance, 2017, Available at SSRN: https://ssrn.com/abstract=2951348 or http://dx.doi.org/10.2139/ssrn.2951348.

Jin, N. & Tsang, E.P.K., Bargaining strategies designed by evolutionary algorithms, *Applied Soft Computing*, 11(8), December 2011, 4701–4712.

Kampouridis, M. & Tsang, E.P.K., Investment opportunities forecasting: Extending the grammar of a GP-based tool, *International Journal of Computational Intelligence Systems*, 5(3), 2012, 530–541.

Kampouridis, M., Alsheddy, A. & Tsang, E.P.K., On the investigation of hyper-heuristics on a financial forecasting problem, *Annals of Mathematics and Artificial Intelligence*, 68(4), 2013, 225–246.

Kern, M., Parameter adaptation in heuristic search – A population-based approach, PhD Thesis, University of Essex, 2005.

Krausz, R. & Gann Treasure, W.D., *Discovered*, Marketplace Books, 2006.

Li, K., Decomposition multi-objective evolutionary optimization: From state-of-the-art to future opportunities, 2021. https://arxiv.org/abs/2108 .09588.

Mandelbrot, B., *The Fractal Geometry of Nature*, W.H. Freeman & Co., 1982.

Mandelbrot, B. & Hudson, R.L., *The (Mis)Behaviour of Markets*, Basic Books, 2004.

Marquez Diez Canedo, J. & Martinez-Jaramillo, S., A network model of systemic risk: Stress testing the banking system 1, *Intelligent Systems in Accounting, Finance and Management*, 16(1–2), 2009, 87–110.

Muthoo, A., *Bargaining Theory with Applications*, Cambridge University Press, 1999.

Rossi, F., van Beek, P. & Walsh, T. (ed.), *Handbook of Constraint Programming*, Elsevier, 2006.

Rubinstein, A., Perfect equilibrium in a bargaining model, *Econometrica*, 50(1), 1982, 97–109.

Russel, S. & Norvig, P., *Artificial Intelligence: A Modern Approach*, Global Edition, Pearson, 2021.

Saini, N. & Saha, S., Multi-objective optimization techniques: A survey of the state-of-the-art and applications, *The European Physical Journal Special Topics*, 230(10), 2021, 2319–2335.

Shleifer, A., *Inefficient Markets: An Introduction to Behavioral Finance*, Oxford University Press, 2000.

Silver, D., Huang, A., Maddison, C.J., Guez, A., Sifre, L., van den Driessche, G., Schrittwieser, J., Antonoglou, I., Panneershelvam, V., Lanctot, M., Dieleman, S., Grewe, D., Nham, J., Kalchbrenner, N., Sutskever, I., Lillicrap, T., Leach, M., Kavukcuoglu, K., Graepel, T. & Hassabis, D., Mastering the game of go with deep neural networks and tree search, *Nature*, 529(7587), 2016, 484–489.

Silver, D., Schrittwieser, J., Simonyan, K., Antonoglou, I., Huang, A., Guez, A., Hubert, T., Lucas Baker, L., Lai, M., Bolton, A., Chen, Y., Lillicrap, T., Hui, F., Sifre, L., van den Driessche, G., Graepel, T. & Hassabis, D., Mastering the game of go without human knowledge, *Nature*, 550(7676), 19 October 2017, 354–359.

Sklarew, A., *Techniques of a Professional Commodity Chart Analyst*, Commodity Research Bureau, 1980.

Tsang, E.P.K., *Foundations of Constraint Satisfaction*, Academic Press, 1993.

Tsang, E.P.K. & Li, J., EDDIE for financial forecasting, Chapter 7. In S.-H. Chen (ed.), *Genetic Algorithms and Genetic Programming in Computational Finance, Kluwer Series in Computational Finance*, Kluwer Academic Publishers, 2002, 161–174.

Tsang, E.P.K., Markose, S. & Er, H., Chance discovery in stock index option and future arbitrage, *New Mathematics and Natural Computation*, World Scientific, 1(3), 2005, 435–447.

Tsang, E.P.K., Olsen, R. & Masry, S., A formalization of double auction market dynamics, *Quantitative Finance*, 13(7), July 2013, 981–988.

Tsang, E.P.K., Directional change for handling tick-to-tick data, *Journal of Chinese Economic and Business Studies*, 2021, https://doi.org/10.1080/14765284.2021.1989883.

Tsang, E.P.K., AI in finance reading list. http://www.bracil.net/teaching/AIF/.

Van Benthem, J.F.A.K., *The Logic of Time*, D. Reidel Publishing Company, 1983.

Zhang, Q. & Li, H., MOEA/D: A multiobjective evolutionary algorithm based on decomposition, *IEEE Transactions on Evolutionary Computation*, 11(6), 2007, 712–731.

Zhang, Q., Li, H., Maringer, D. & Tsang, E.P.K., MOEA/D with NBI-style Tchebycheff approach for Portfolio management, Proceedings, Congress on Evolutionary Computation, Barcelona, Spain, 18–23 July 2010, 3008–3015.

INDEX

Note: Page locators followed by 'n' refer to notes.

Printed in the United States
by Baker & Taylor Publisher Services

Printed in the United States
by Baker & Taylor Publisher Services